# MANDATED SCIENCE:

# SCIENCE AND SCIENTISTS IN THE MAKING OF STANDARDS

ENVIRONMENTAL ETHICS AND SCIENCE POLICY

*Editor:*

K. S. SHRADER-FRECHETTE

*Department of Philosophy, University of South Florida, Tampa, U.S.A.*

# MANDATED SCIENCE: SCIENCE AND SCIENTISTS IN THE MAKING OF STANDARDS

LIORA SALTER

*Department of Communication, Simon Fraser University*

*with the assistance of Edwin Levy and William Leiss*

KLUWER ACADEMIC PUBLISHERS

DORDRECHT / BOSTON / LONDON

**Library of Congress Cataloging-in-Publication Data** 

Salter, Liori.
Mandated science.

Bibliography: p.
Includes index.
1. Environmental health—Standards—Case studies.
2. Environmental health—Government policy—Decision making—Case studies.
I. Leiss, William, 1939– II. Levy, Edwin. III. Title.
RA566.S25 1988 363.1´062 88–6934
ISBN 1–5560–8057–3
ISBN 1–55608–077–8 (pbk.)

---

Published by Kluwer Academic Publishers,
P.O. Box 17, 3300 AA Dordrecht, Holland.

Kluwer Academic Publishers incorporates the publishing programmes of
D. Reidel, Martinus Nijhoff, Dr W. Junk and MTP Press.

Sold and distributed in the U.S.A. and Canada
by Kluwer Academic Publishers,
101 Philip Drive, Norwell, MA 02061, U.S.A.

In all other countries, sold and distributed
by Kluwer Academic Publishers Group,
P.O. Box 322, 3300 AH Dordrecht, Holland.



Printed in the Netherlands.

# Table of Contents

**Preface** ix

**Chapter One: Introduction to Mandated Science** 1
Identifying Mandated Science 1
The Character of Mandated Science 5
The Approach to be Taken in the Study of Mandated Science 10
Standard Setting: A Case Study of Mandated Science 12
The Design of the Study 14
Specific Methodological Decisions 17
The Organization of the Book 19

**Chapter Two: An Introduction to Standards** 20
The Features of Standards 21
Confusions in Terminology 26
The Data Problem in Standard Setting 28
The Debates about Standards 31
Standard Setting as an Example of Mandated Science 34

**Chapter Three: In the Eye of the Storm**
*Case Study One: The American Conference of Governmental Industrial Hygienists* 36
The Early History 37
The Active Phase 39
The Transition Period 40
ACGIH Today 43
Membership of the TLV Committee 44
Standard Setting in ACGIH 47
ACGIH Standards 49
Controversies about Standards 50
The Status of ACGIH Standards 51
The Use of ACGIH Standards 53
Discussion 55

**Chapter Four: Alphabet Soup**
*Case Study Two: The Codex Committee on Pesticide Residues* 67
The Codex Alimentarius 68
The Codex Committee on Pesticide Residues (CCPR) 73
The Joint Management Committee on Pesticide Residues 76
The Three Organizations 79
The Standards 82
The Status of Codex Standards 85
Discussion 88

**Chapter Five: Political Chemicals**
*Case Study Three: The Toronto Lead Controversy* 98
Background Information 99
Standards in the Toronto Lead Controversy 103
The Toronto Lead Controversy (1) – Early History 104
The Toronto Lead Controversy (2) – the Case Goes to Court 109
The Toronto Lead Controversy (3) – Words Become Dangerous 113
The Toronto Lead Controversy (4) – Studying the Problem 116
The Toronto Lead Controversy (5) – The Hearing Acts as a Court 119
Discussion 122

**Chapter Six: An Economic Poison**
*Case Study Four: Pentachlorophenol* 132
Some Background Information 133
The Standards 137
History of the Controversy 139
Discussion 154

**Chapter Seven: Standards Revisited** 160
The Characteristics of Standards 160
The Character of Standard Setting: The Two Organizations 162
The Character of Standard Setting: The Two Controversies 165
Standards and the Debate about Regulation 178
The Debate about Standards: Prescriptive versus Performance Standards 180
Standards and Mandated Science 183

**Chapter Eight: Mandated Science** 186
The Character of Mandated Science 187
Questions Arising from the Study of Mandated Science 190
The Debates in Mandated Science 202
Conclusions from the Study of Mandated Science 206

**Notes** 210

**Index** 219

# Preface

For a long time I would not eat strawberries. In 1977, a scandal broke about a testing laboratory having falsified the data that was used to register a large number of pesticides. The Canadian government, along with several others, began the process of re-evaluating both the procedures for testing and these specific chemicals. One chemical proved particularly controversial, the commonly-used pesticide named captan. In light of the controversy, which was manifest in a conflict between two government departments, in 1981, the Canadian government chose to appoint a special panel of experts to advise them. I was a member of this expert committee.

The experience on the captan committee did little to reassure me, either about captan or about the way that decisions had been made about many pesticides in widespread use. Although it could not be demonstrated that captan was dangerous to people in the amounts to which they would likely be exposed, the animal studies provided the basis for concern. Prudence required at the very least that consumers take the precaution of washing their fruit, for captan is widely used on apples, cherries and berry fruits. Captan residues wash off apples relatively easily; they are less easily removed from berry fruits, such as strawberries.

I had images of my children when they were young, munching their way through the strawberry fields. I now knew about the difficulty of controlling when the pick-your-own farms let people into their fields after spraying. In a classic after-the-fact fashion, I reasoned that a little caution, even too late, might fend off any dreadful consequences. I now eat strawberries – thoroughly washed, of course – but the study of mandated science has been as troubling as my hindsight about my children eating from the pick-your-own strawberry fields.

I suspect I am typical of many consumers. I had assumed that farmers could make decisions about whether to use pesticides, and that most foods I ate were free of residues. When we began this study, I had little awareness of either the number of pesticides in use or the number of times any specific fruit or vegtable might be sprayed. I was shocked to find out that an apple might be sprayed more than twenty times, and with more than one pesticide, before I bought it.

I do not assume that all pesticide residues are dangerous, and I am willing to accept some risk for the pleasure of easily available and unblemished fruit. What is troubling is the realization of how completely the use of pesticides is integrated into the production of food, and how little choice commercial farmers actually have. It is useful for me to think of chemical use in food production in the same way as mechanization in manufacturing. Indeed, it is reasonable to consider pesticides simply as chemical mechanization, and as inevitable as factory mechanization in industrial and industrializing societies.

Personally, I wish it were not so. I wish it were possible to control the use of chemicals in food production quite strictly, and to avoid them in many situations. I now believe that such control demands a radical transformation of the process by which food is produced, and of the social and economic relations in food and food-related industries. It demands fundamental changes in land tenure and use, for many times the agricultural land would be required if chemical mechanization were to be abandoned in favour of other farming methods. I am sympathetic to the demand that these changes take place, but I think it is naive to underestimate the magnitude of the revolution involved in turning away from large-scale chemical mechanization.

In the course of our study on mandated science, I also became aware of the degree to which chemicals are used in industrial production, and the magnitude of any decision to exert strict controls, even with respect to hazardous waste. It is hard for me as a citizen to comprehend that there are groups with a significant vested interest in the production of hazardous waste, and similarly with an interest in preventing its control. Given the enormous costs of even assessing the potential dangers of industrial chemicals, I am genuinely surprised that so few serious accidents have occurred, although it is a mistake to underestimate the less easily demonstrable damage that chemical exposure has caused.

Sobering thoughts. For some people, the answer is to advocate a less chemically-dependent world, regardless of the radical nature of their demand; for others, it is to call attention to the crises and accidents that do occur as a means of awakening citizens to the nature of the situation. For me, it is has been to turn my attention to the way that decisions are made – or not made – about chemical and pesticide control. Reference is often made to "standards". Chemical exposure is said to be under control if chemical pollution or exposure meets an accepted standard. I want to know where these standards come from and whether there is any solid basis for trusting in them. This is my motivation in writing this book, although strictly speaking, it is not its origin.

What I have found is not particularly reassuring. More important, I have found that the origin of standards is not widely known or researched, and that even my learned colleagues who write extensively (often with similar motivations) about the enforcement of standards have little knowledge about their origins. Before one can take any platform to debate the wisdom of standards, it is necessary to identify how they come into being. Thus, this book is not a polemic, and no statements have been made within it about the dangers of chemical mechanization, about the adequacy of any particular standard or about the toxicity of any chemical. It has another purpose, which is simply to lay bare the kinds of decisions and decision making processes that result in standards or their revision. I believe this task must be done first.

The book itself has a different origin. It is a product – one of several – of an extended collaborative study of science and public policy. Ed Levy, one of the researchers has written extensively on science and social and political institutions. Bill Leiss has written on technology and human needs. I have done a major study on public inquiries, focussing primarily on the role that scientists and members of the public played in these inquiries, and considerable work on regulation. We came together to study standards as an example of how science, values and public policy were combined in the origination of standards.

The term "mandated science" was coined by Andrew Thompson, the Director of the Westwater Institute at the University of British Columbia. Like us, he was concerned with the way in which the policy "mandate" affects the kind of scientific assessment that is done, primarily but not exclusively in environmental regulation. We originally planned a joint project with him, but have gone in different directions with our research. Nonetheless, this study owes a debt to his original concept, which we have extended and developed as an integral part of our research.

Our research involved six case studies. Ed Levy took primary responsibility for the research on pentachlorophenol and fluoride. Ed and I jointly researched ACGIH. Bill Leiss and I worked together on the Toronto lead controversy. I took primary responsibility for the captan and for Codex, and was project and research director throughout the three years of the research. We drew upon the assistance of many people, but one deserves particular attention. Rory Daniel was the research assistant for four of the studies, and her careful scholarship made a substantial contribution to the final product. Other researchers include: Linda Drury, Barbara Nelson Hughes, Richard Pinet, Johanne Gregory, Janice Peck, Ron Trepanier, Richard Palidwor and Debra Pentacost, Peter Campbell, Lance Cooke, Rick Gordon, Jeff Hirschfield, Duncan MacIntyre, Diane Pennington and Simon Thacker. We are indebted to them for their dedication, interest and excellent work. We would also like to acknowledge the support of the Social Science and Humanities Research Council of Canada during the three year period of the research.

As I have mentioned, this is one, but not the only product of our study of mandated science. Bill Leiss has now conducted several evaluations for the Federal Government of decision making procedures relating to environmental and pesticide issues. He has written several articles on risk communication, and is actively involved in developing protocols for risk assessment in Canada. Ed Levy has elaborated the study of the scientific controversy over pentachlorophenol, examining the experience of scientists in that controversy and issues related to the studies they conducted. In addition, he has drawn upon this study to further his continuing work on scientific and value judgements. I have published the study of captan in another context, and written on the peer review process used in standard setting in Canada. Together with other collaborators, we have embarked on a new study, this one entitled "The Discourse on Risk". The new study examines how the concept of "risk" has emerged as significant in a variety of different disciplines, and its impact within the public policy debate.

Finally I should mention a number of friends and colleagues whose support has been important to us. We held two workshops during the study of mandated science, and drew upon the lively exchange among ourselves and Trevor Pinch, Brian Wynne, Andrew Dorcey, Andrew Thompson, Sheila Jasanoff, Claire Franklin, Frank Cedar, Judith Miller, Kim Roberts, Steve Straker, Gus Brannigan and Bob Anderson. Without the support and guidance of Julia, Benjamin, Jennifer, Marilyn, Rick, April, Colin, Ammon, Steve, Clayton, Harriet, Emma and Kate neither our research nor its several products would have been possible. Let no one underestimate their contribution. To all these people, and to the more than two hundred people we interviewed, we extend our thanks.

*Liora Salter,*
*May l987*

Chapter One

# Introduction to Mandated Science

We regularly use science and scientists in the making of public policy. We depend on scientists, for example, to tell us whether we should be worried about radiation and whether nuclear power plants are safe. We expect science to tell us if a pesticide is likely to cause disease, so we can decide whether and under what conditions to permit its use. Public confidence in medicines, clear air and water, safe and quiet working conditions and the reliability of some products rests on the belief that scientists have been consulted about their safety. Our image of scientists pictures them at work in the laboratory; we seldom raise the question of how scientific information moves from the laboratory to the world of politics and policy making.

This book is about the role of science and scientists in making public policy. The term "mandated science" is used to refer to this role. We have set about to examine how science makes the transition from one context – the world of scientific work – to another – the daily practice of political life. We do not assume that all science takes place in laboratory settings, for it has become obvious to us that much of the scientific work used in policy making does not come directly from the laboratory. Along the way – from laboratory to decision making – other things happen that involve science and scientists but that conform to few of our popular images about scientific work.

We do not think of these things as the corruption of science by politics. Nonetheless, we have found that science and scientists in the realm of policy making are acting in a special world. It is a world with definable characteristics, pressures and constraints. There is a need to understand this world, if the public is to have full confidence in decisions taken on its behalf about pollution, industrial health or safety in food. We use the phrase "mandated science" to describe this world, as well as the role of scientists within it, and our need is to understand how the pressures and constraints of mandated science are felt by scientists and governments alike.

## Identifying Mandated Science

The term "mandated science" refers to the science that is used for the purposes of making public policy. Science, here, includes the studies commissioned by government officials and regulators to aid in their decision making. This scientific work is designed and carried out solely for the purpose of supporting particular regulatory decisions. It also includes scientific work originally produced in more conventional scientific settings. This scientific work is indistinguishable on the printed page from an academically-oriented report. It becomes "mandated" when an individual study is evaluated in terms of the conclusions it can offer to policy makers about the merit of particular regulations.

Let us take an example. Governments often appoint expert committees to aid their work in deciding whether new regulations are required – say, an expert committee asked to make recommendations about a potentially dangerous pesticide. It is widely known that the public is concerned about the potential danger to its health posed by widespread use of this pesticide on food crops. The pesticide has been in use for many years, however, and there have been no documented complaints. Nonetheless, several studies, reported in publications such as *Science* or the *New York Times,* suggest there is cause for serious concern about its health effects. Some rat studies have shown that the pesticide has the potential to create cancers in laboratory animals. The regulatory status of the pesticide in question has become controversial, in other words. The job of the expert committee is to provide government with some advice, and in doing so, to help resolve the controversy.

The committee holds its initial meeting and reviews the situation briefly. It determines that a literature review should be done, and later meets to discuss the implications of various studies in terms of the directions they suggest for public policy. None of the studies being reviewed contains such directions or recommendations, of course. These studies are the kind of scientific work that is normally published in the academic literature, where discussion of the moral or policy implications of the study would seem highly inappropriate. The committee must draw out these apparently non-scientific implications for itself, reading between the lines if necessary to create the sense of certainty about what has been found in the laboratory studies on the pesticide.

The committee also has access to a body of literature that is not published in normal academic journals, nor always subject to a process of peer review. These are the studies commissioned, and often carried out by the companies manufacturing the pesticide in question and seeking regulatory approval. These studies are similar to those in the academic literature, in that they follow conventional scientific methodologies, and are conducted by persons with scientific credentials. These studies are distinguishable from the academic literature, in that it is relatively easy to draw conclusions from the research to support particular regulatory decisions. In the design of these studies, it seems, attention has been paid to narrowing the scope of the subject to be investigated so that the conclusions will be readily applicable to regulatory decision making. Both types of scientific work are included in the definition of mandated science.

Our committee is not satisfied with either of these bodies of scientific literature. Some questions have not been investigated in that literature, and several existing studies have produced contradictory findings. This committee is an exceptional one, for it has resources to allocate to original research. It commissions two types of new research. The first is called a short term feeding study, and it examines the effect of the pesticide on a new group of laboratory animals. The second is a review of a particular body of scientific literature, a task for which the committee members themselves lack expertise and resources.

In commissioning these studies, the committee is very clear about its needs. The research should meet the test of good scientific work – although the peer review in this case will be through the committee itself – but the researchers understand from the start that the committee requires results that are useful for the development of regulations. Whether the researchers can meet this requirement remains an open question, but some self-selection does occur. Only scientists who are comfortable with the pressures imposed by an expert committee working for government are likely to undertake the work. The work of the committee and the research it commissions are also mandated science.

Mandated science is all of these examples of research, plus the work of the committee in evaluating them. The term "mandated" refers most particularly to the pressure being placed upon science and scientists to reach conclusions that can lead to public policy or government regulations. The term "mandated" implies an argument that will be made explicit in this book. We believe that a mandate to develop recommendations or decisions for public policy exerts a pressure on science that is reflected both in the activities of scientists and in their work or its interpretation.

We would find easy agreement with this argument from those scientists who testify regularly before regulatory tribunals and the courts. We think, however, that even conventional science can be altered – in effect, after its publication – by its use for the purposes of public policy. Although it is true that the scientists who publish in academic journals usually pay little attention to questions of public policy – except in so far as the funding for their continued research depends upon the usefulness of their work for policy purposes – their work constitutes a resource for those making such decisions. These more academically-inclined authors might well be surprised – although not necessarily dismayed – by the way their work is used by governments. They might be surprised, for example, by the degree of certainty that is attached to their conclusions or by the implications drawn from their work by expert committees. Their studies, if not they themselves, become part of the world of mandated science when their research is used and intepreted in that world.

Our second argument is that mandated science is a type of scientific activity with particular characteristics. In other words, the scientific activity of an expert committee, of the scientists who testify for regulators, of the regulators or courts that assess scientific information and the studies that are used for the purposes of making public policy all constitute a world of science that is best described in its own terms. These terms are derived from the fact that all this activity takes place under the auspices of a mandate to produce recommendations for government. We have set out to investigate the character of mandated science through a series of case studies.

Our purpose is not to compare mandated science to what we might call the "conventional" modes of scientific activity, for we have not studied conventional science. It may be that the characteristics of mandated science that we have uncovered in our research are indeed similar to those of conventional science or, alternatively, that mandated and conventional laboratory science differ significantly. We do not know. We refer to the image of laboratory science, and to perceptions of "good science" only because these perceptions are often discussed in mandated science. We use mandated science as our frame of reference because it focusses attention on a series of questions that have important public consequences.

We want to know what influence the policy mandate has upon the way scientific work will be conducted or understood. We are concerned about how well science is – and can be – incorporated into the regulatory process. Are public hearings a valuable and effective debating arena for conflicting interpretations of scientific literature? Do the courts or expert committees provide good settings for the airing and resolution of scientific questions? Why are so few scientists involved in regulatory work and why is a high degree of frustration expressed among many who are?

Mandated science must be understood as a separate sphere of scientific work if these questions are to be answered adequately. For increasingly, decision makers and their publics are placed in a quandary. On one hand, they are increasingly dependent upon science and scientists in a world characterised by complex technologically based economies. On the other hand, it is increasingly apparent that science cannot provide the clear answers that governments seek, at least not at the time when regulatory decisions are required. Moreover, science often provides conflicting answers that cannot be interpreted exclusively by reference to scientific norms.

For example, because of the cost of research, someone must make a choice about which studies to commission and how extensively to review the scientific literature. Someone must decide how to interpret studies whose findings are uncertain, ambiguous or not directly applicable to the decision to be made. Someone must decide where the burden of proof lies with respect to health and environmental hazards: are chemicals considered safe until proven otherwise or should they be considered hazardous until evidence indicates that they can be used safely. The answers that scientists provide involve judgements which are intrinsically related to science, but which are not themselves scientific in the conventional sense of the term.

This quandary – the increasing dependence upon science and technology and an increasing concern or skepticism about the kind of answers that science itself can provide – is the heart of the matter. As a consequence, the study of mandated science is both a practical exercise designed to provide answers about the best way of handling scientific information in the policy process and, at the same time, an inquiry about the status of science in public life.

## The Character of Mandated Science

Based on the case studies in this book, we will argue that mandated science has four characteristics that affect both the practical resolution of problems of incorporating science in policy decisions and the status of scientific work as a human endeavor. Each poses a paradox.

*Mandated Science as an Idealized Science:* We will suggest that the first characteristic of mandated science arises from the fact that it takes place in public view – indeed, it is subject to a particularly imposing form of public scrutiny. Mandated science has as its primary audience the governmental body that has mandated the work or seeks recommendations from it. Because governmental bodies must justify their actions in the political process, the science they seek is one that is capable of being justified and explained to a wide variety of publics and interest groups. This science must be intelligible to the non-scientific audience. It must facilitate clear choices. It must represent a body of evidence upon which decisions can rest and be seen to be rational. In mandated science, science is used to support public policy.

We have come to believe that only an ideal science could serve these non-scientific purposes, and one of the features of mandated science is its reliance upon an image of an ideal science. This ideal science has several features. One is an assertion that science is value free, and that scientific conclusions are independent of the use to which they are put. Second, mandated science relies upon the assertion that the scientific method will produce invariably credible results. And third, and most important, science is depicted ideally as an inherently public enterprise, an activity that is produced and vetted through open debate, anonymous peer review and academic publication. These three assertions are necessary if science is used to justify government decisions.

A deep paradox apparent from our research is that mandated science conforms to none of these ideals. First, in mandated science, the moral dilemmas posed by scientific knowledge are made explicit, and few assume that science is fully independent. Even if one were to argue that laboratory science is value-free (which we do not), mandated science always involves a close, and often acknowledged relationship between science and values.

Second, we have observed that in mandated science the conventional methodologies of science often produce conflicting reports that *cannot* be resolved by further properly conducted studies. In mandated science, scientific studies appear to be lined up on either side of a public debate about regulation, and it is a commonplace observation that "science can be bought", or at least that evidence can be marshalled for any value position in the regulatory debate. In mandated science, reference to an accepted scientific methodology is not sufficient to establish the credibility of a research study.

Finally, a proportion of mandated science is neither peer reviewed nor published, nor is it part of an open literature. Whether or not it meets the standards of conventional science, we will argue that mandated science cannot be understood as an intrinsically public enterprise. The scientific debate in some regulatory tribunals is often more democratic than that which occurs in academic journals. Nonetheless, the fact that many of the scientific studies used in mandated science are not published, and that much of the information being reviewed by regulators is propietary, undermines the contention that mandated science is either public or democratic in nature.

Our case studies provide the basis for the conclusion that mandated science is science that incorporates an idealistic view of the scientific enterprise. At the same time, it is a type of scientific activity in which the discrepancy between the ideal and the actual activity is significant, and easy to identify.

*Mandated Science as Science Imbued with Legal Considerations:* It is not particularly remarkable that the decisions taken in the context of mandated science are imbued with interests of an economic and social nature. What is worthy of note is that these decisions, and thus the debates leading up to them, are also deeply infused with legal issues that compound their economic and social implications many times over. How could it be otherwise? A decision to remove a pesticide from the market will invariably be challenged in a court of law, somewhere. Moreover, any damage resulting from the use of the chemical is likely to result in some court action.

Our research suggests that the second characteristic of mandated science is its complex and symbiotic relationship with the legal process. Simply put, decisions taken on the basis of scientific evidence by policy makers and regulatory bodies have significant legal implications. These decisions are the result of discussions about liability for harm, and about measures that might be taken to restrict both harm and limit liability for it. They result in a greater or lesser capacity to take court action for possible negligence or damage to human health and the environment. The actual science – the reports and conclusions – of mandated scientists are used as legal evidence, and thus are subject to the standard of proof required by the courts. These reports are part of a strategic arsenal of legal evidence that can and will be marshalled in case of a legal proceeding. [1]

The scientists who enjoin the debate in mandated science act – as they do in most other contexts – as if legal questions had little influence on their scientific conclusions. Indeed, we believe that they must act blindly with respect to the legal implications of their scientific work in order to maintain their credibility as scientists. This, then, is the second paradox of mandated science. To be considered as good science, scientific information must be developed and presented without regard for its legal implications. Yet the testimony given by experienced participants in mandated science is shaped by an awareness of legal standards of proof and by knowledge about how to use scientific information effectively in a court or regulatory setting.

We have found that scientists compromise their claim to independence to the extent that they openly recognize the legal constraints within which they operate in mandated science. At the same time, these scientists must function in a legally informed manner (and indeed, reach conclusions that can be tested according to legal standards of proof) if their work is to be valuable to the legally-oriented bodies that mandate or use their scientific work.

Mandated science involves legal issues, and a scientific debate that is judged by the norms of both science and the courts. Our research suggests that scientists who are experienced in mandated science often proclaim their allegiance to scientific norms, as they must to maintain their scientific credibility, but they are equally committed to and knowledgeable about the demands of the legal process. The two types of norms, scientific and legal, are not easily combined, and indeed we will argue, mandated science has its own norms and conventions that are quite different from those of either the scientific or the legal process.

*The Character of Debates within Mandated Science:* The third characteristic of mandated science is the unique nature of discussion and debate within it. Scientific debate in mandated science is clearly recognizable as neither scientific nor legal nor public policy debate – rather, it has its own style, methods of argumentation and uses of language.

Some illustrations from our research about the nature of the debate in mandated science will be helpful to introduce the perspective taken in this book. We have found that scientific information is presented in mandated science as if it were evidence in a public policy debate infused with interest group considerations. Emphasis is placed on closure, on the "bottom-line" results of what can be learned from any study. Our research suggests that no experienced participant in mandated science will muse openly and voluntarily about contradictory findings, about scientific uncertainty or about the methodological judgements that were made in the conduct of the research, unless such musings constitute part of the evidence being submitted.

Questions of fairness and democratic rights arise in mandated science, and scientific conclusions will often be justified as reflecting a consensus of interested parties or a democratic process of decision making. Both laypersons and scientists take part in scientific discussions. Our research indicates that scientists often give "scientific testimony" on matters such as regulation that are not, strictly speaking, related to their own expertise. Even when they do not, their scientific contribution may be understood as part of an advocate or political argument. Finally, we have found that in mandated science the emphasis is on the evaluation of science rather than on its conduct.

The language, the use of particular words and phrases, of mandated science reflects the peculiar character of its debate. We will suggest that words or phrases lend themselves to being battlegrounds for conflicting interpretations of science that benefit some political or interest groups and not others. The term "scientific uncertainty" is a good case in point. In a scientific context, reference to "scientific uncertainty" is unexceptional. In our study of mandated science, we found that reference to "scientific uncertainty" had important political consequences. The credibility and usefulness of a study could be challenged by a lawyer in cross-examination simply by reference to the scientific uncertainty attached to its conclusions. Regulators could justify action or inaction also by referring to scientific uncertainty.

This poses what we suggest is the third paradox of mandated science: In order to maintain their credibility as scientists, participants in mandated science must adhere closely to conventions of scientific debate that are acceptable to other scientists. They must speak as if they were speaking with other scientists. To be effective in the policy arena, however, these same scientists are often also required to do otherwise. They must speak with an awareness that others – whose preoccupations and interests are quite different – will use what they say to further goals that are unrelated to science.[2] In effect, we have come to believe that scientists who regularly give expert testimony for regulators or for expert committees speak to at least two different audiences simultaneously, and know that their words can be subject to conflicting interpretations by each.

In sum, we will use a metaphor to argue that language is transparent in science, as it is in most fields of endeavor. That is, the words or phrases themselves are not subject to openly conflicting meanings. We will suggest that the language in mandated science is opaque, in the sense that the language – including scientific terms – becomes a battleground not only for conflicting interpretations of scientific information but also for legal and interest group conflicts. In mandated science, a scientist addresses many different audiences simultaneously, and the debates in mandated science about scientific information reflect this situation.

*Mandated Science and Moral Issues*. Finally, we will suggest that the fourth characteristic of mandated science is that it makes explicit the moral dilemmas posed by science. The members of our hypothetical expert committee were chosen because of their scientific expertise. Yet, the advice they must give is about how a pesticide should be regulated. The latter is a different question from the former, and their expertise might not be of much assistance in answering it.

The committee members, and also the scientists they consult, must extrapolate from what is established in the scientific literature, draw out implications from particular studies, and bring together scientific material from different disciplines to reach a single decision. They make other kinds of decisions in the process. The members of an expert committee can decide to err on the side of caution or not. They take economic and social questions into account in varying degrees considering, for example, the effect on the farmer's livelihood of withdrawing a pesticide from use.

What they, and any other body with both a scientific and policy mandate, must do – because of the nature of that mandate – is make choices. The intention is that their choices will be informed by scientific understanding and that their interpretation of the scientific literature will be sensitive to the norms of science and its particular limitations. Nonetheless, we believe that in accepting their task, the members of the expert committee agree to recognize constraints that scientists seldom recognize or acknowledge explicitly, and to relax constraints that scientists publicly claim to abide by. They agree to recognize the implications for society of the conclusions they draw from scientific data. They agree to consider moral questions, at least obliquely. And they agree, in most cases, to go beyond the normal activities of science in translating scientific conclusions into recommendations for policy.

This creates the fourth paradox, for the members of an expert committee are chosen precisely because they are insulated from the pressures of interest groups. They are expected to act as if they were neutral arbiters – as expert scientists – of the issues being debated. They are chosen, in part, because science is seen to be value-free, and their work, as scientists, is seen to reflect an unbiased view. We have found that because they are working in the sphere of mandated science, they conduct assessments that are only partly scientific, and make decisions that have direct political and moral consequences. They face the problem that a statement about scientific issues is inherently also a statement which privileges some interests and not others.

Our members of the expert committee were chosen, then, because they could rely upon their scientific training to render themselves relatively free of interest group and moral constraints. The work they do, however, is mandated science, in which moral and political choices are made regularly. Our committee members might choose to ignore the effect of their report on the various groups with an interest in pesticides and speak informally about the awkward lobbying efforts of some industry or environmental group. But none are so willfully ignorant as to suggest that their final document – at the very least – is free from similar constraints. They know, for example, that banning a chemical affects a $500 million industry, while recommending its continued use could have serious health effects that become evident only ten years later. We will argue that this knowledge affects their recommendations as much as does their knowledge about the characteristics of the natural phenomenon. Again, we think it could not be otherwise.

Our argument in this book is that these four characteristics – the idealization of science, the legal substratum of scientific debates, the peculiar nature of the debate itself and the manifest interplay of science and values – distinguish mandated science from at least the conventional view of science. We believe these characteristics are a result of the imposition of a mandate that is derived from the needs of policy, not science.

We will make no claim that other kinds of scientific work lack mandates, or indeed that science generally should be conceived only as a knowledge-seeking enterprise, isolated from social and political influences. And we make no claim that the practices of laboratory science conform to the conventional picture of the scientific enterprise. On the basis of our research, we can say that mandated science does not conform to the conventional picture of science as a knowledge-seeking exercise nor of the ideal science that is often discussed in the context of mandated science. However much policy or other mandates or missions affect other scientific work, we believe that the character of mandated science, as we have observed it in six case studies, comes from the close, entangled and ambiguous relationship this particular activity has with science, values, law and public policy.

## The Approach to be Taken in the Study of Mandated Science

Our argument in this book – that mandated science can be distinguished as a discrete form of activity – runs contrary to much of the standard wisdom about the relationship among science, values and public policy. A rationalist view of this relationship is firmly entrenched within the literature. In a now classical discussion of risk for example, Lowrance argues that the process of scientific assessment (the scope and severity of risk - risk assessment) can be clearly distinguished from that of deciding how risks should be regulated (risk management).[3]

Following his model, others have suggested that the two spheres of action – science and policy making – can be brought together in the design of a regulatory process that separates scientific and values issues for separate and sequential consideration.[4] The American regulatory process for pesticides is – in theory at least – an example of such a two step process, as indeed is the approach taken in Germany and a number of other western industrial countries. [5] Some of the earlier literature on the idea of establishing a science court falls within this approach, for in it the argument is made that a separate tribunal should be established to resolve scientific controversies before political decisions are made.[6]

While we will have something to say about the design of institutions that make policies with scientific implications, it is our contention that the rationalist model oversimplifies and distorts the picture, even to the detriment of practical regulatory planning. At the very least, if we accept the contention that mandated science has its own characteristics, then we cannot simply link science and policy through that model, as if the linkage posed no problems.

We have already suggested that the language used in mandated science is not transparent. Here we extend the point. We will argue that the activities of mandated science are not transparent. They have a significant effect upon scientific debate and, indeed, perhaps even on the content of scientific information itself. They also have an effect upon the scope and nature of the value debate about public policy options. At the very least, then, the existence of a sphere of mandated science between conventional science and policy making confuses the picture. We think it does more. We think it alters the relationships between science, values and public policy in very significant ways. Our case studies are illustrations of *how* the relationship between science, values and public policy are shaped in the sphere of mandated science, and they provide the reasons for rejecting the rationalist view.

We also take issue with a different school of thought in the literature about science, a school best characterized by the ethnomethodological approach of its practitioners. Laboratory science does not, it appears on examination by Latour and Woolgar[7], Knorr–Cetina[8] and their colleagues[9] conform very closely to conventionally accepted scientific norms, nor is the highly regarded scientific method practiced in easily recognizable ways within a laboratory setting. Rather, science can be viewed as a set of social practices that warrant particular forms and outcomes of knowledge.[10] At least in part, scientific knowledge is a product of a negotiation within the scientific community and through its various legitimating institutions.[11]

Our objection to this approach, to which we in fact owe a great deal, is a moral one. We begin with the moral position that it *does matter* which chemicals are carcinogens and that some chemicals are more dangerous to human health than others. These are matters to which we think science can make a contribution, more or less adequately for the purposes of public policy. Viewing science simply as a series of social practices appears to us to dissolve the importance of the knowledge that can be gained in a laboratory or other scientific settings. At worst, one can be left with a nihilistic, relativist theory of human knowledge – scientific knowledge is simply that which has passed through the various conventional practices and negotiations of science and which has been legitimated as such.[12]

The study of mandated science does not begin with an ideal of the scientific enterprise. For this stance, we are indebted to the sociologists of science who have identified many social practices of science. Nonetheless, our study does begin with an assertion that knowledge about natural phenomena, collected and analysed through the normal practices of scientific work, has a crucial role to play in protecting human health and the environment. Thus, we have delimited our discussion about the nature of science generally, and we view science primarily as *a strategy* for making contributions to matters such as the evaluation of chemical hazards. It is, we believe, a particularly useful strategy, given what we want to know. This is the "science" that we understand when we talk about the relationship between science, values and public policy.

## Standard Setting: A Case Study of Mandated Science

Although the term "mandated science" is a new one, there are other studies of the same phenomena. Most deal with the way in which hazards are evaluated. Often the emphasis is on determining an appropriate method for risk evaluation, and different institutional approaches are compared. There is, for example, a developing literature on risk assessment and risk management, some useful studies comparing cancer policies and chemical evaluations procedures in different countries, and a number of case studies of controversies about the regulatory status of particular chemicals or projects.[13] Furthermore, there has been an interesting debate about the nature of the scientific problems posed in mandated science.[14]

We have taken a somewhat different approach, choosing to study standard setting, rather than either science itself, or risk evaluation, or decision making procedures or particular chemical controversies. This book deals with the standards for potentially harmful chemicals in the workplace and environment. It includes an examination of voluntary, industrial and regulatory standards as they have been developed in a number of different countries. Our purpose is to examine the general process of creating environmental standards, rather than specific standards, and to use the origination of standards as our example of mandated science.

We think that standard setting is a good example of mandated science. We will suggest that the numbers – the standards – are seen to be the scientific component of the regulatory process. Often, we found that the origin of these numbers is taken for granted. Regulatory decisions are concerned with whether the numbers should be adjusted, and how they should be applied. In tracing the origin of the numbers, we are dealing with many assumptions about science and about the role of scientists in making public policy. In fact, we will argue, in the setting of standards, the role of science and scientists is ambiguous and standards have a complex and varied regulatory status.

Three aspects of standards make standard setting particularly appropriate as a case study of mandated science. First, it is easily observed that decisions about standards reflect the contradictory demands of science and of economic interests. Scientific questions about the level of pesticide residues on onions, for example, are important because high levels of pesticide residues often have negative consequences for human health. But the standard set for the amount of pesticide residue on an onion will determine who can import onions and the options for their export. Standards function in trade like the gates on a pinball machine. They direct the flow of trade, but they are not themselves the purpose of the exercise. A study of standards provides an excellent opportunity for examining the relationship between science and economic interests.

Secondly, the term "standard" is a very good example of how words assume the status of a battleground between interest groups in mandated science. Standards can refer to both maximum and minimum levels of acceptable harm. The term "high standards" implies a search for excellence. But environmental standards do not mean maximum possible environmental protection, but rather the minimum levels of acceptable performance with respect to the creation of environmental pollutants. The term "standard" can also mean standardization or uniformity. In environmental standard setting, this standardization is of great importance, since standardization (or lack of it) is what influences trade and commerce.

We have found that the participants in mandated science know about the ambiguities in the term "standards". They use the conflicting meanings of the term "standards" strategically, to support their various arguments in the policy debate. Sometimes, reference to standards implies that some form of scientific assessment has occurred. Sometimes it does not. The ambiguity of the language, and the implications of this ambiguity for science, are at least accepted, if not promoted in the standard setting process. If we believe that an examination of language is important for the study of mandated science, then standard setting provides an excellent case study.

Finally, standards are the product of negotiations that occur in a variety of settings and jurisdictions. Our research suggests that individual standards move up and down (become more or less stringent) depending upon who is influential at the time, upon national and international priorities as well as in response to new developments in scientific knowledge. We will argue that the "fungibility" of science with respect to local, national and international politics is a critical component of mandated science. In the case of standards, it is easy to demonstrate not only the interpenetration of scientific and economic issues, but also the legal and political underpinnings of the negotiations that take place about particular standards.

The use of standard setting as the example of mandated science permits us to focus on the relationships between science and economics, between science and legal relationships, and on the use of language. These are important relationships that are often obscured in a study of how particular countries evaluate chemicals, or in providing case histories of controversies about specific standards. It also permits us to study mandated science paying little attention to the differences among countries, for the same standards are used in many countries with different standard setting procedures.

Several problems occur because of our choice of standard setting as the example of mandated science, however. First, we have observed that the term "standard" is used differently by various groups and in different countries, and standards are often called by different names. The terminological confusion is not easily sorted out, nor is there an adequate guide in the literature. It will be necessary to examine the terminology, and to arrive at a definition of the phenomena that takes so many different forms, even if this examination appears to divert us from the study of mandated science.

Secondly, there are three current public debates about standards. The first debate is about the merit of regulatory, as opposed to voluntary standards. The second is also about the usefulness of performance, as opposed to prescriptive standards. Both of these debates reflect a preoccupation with deregulation, but this is not a book about either government intervention or deregulation. Nonetheless, it will be necessary to examine these two current public debate about standards, because they highlight some interesting aspects of the relationships among science, values and public policy. The third debate is about methods of enforcing standards. We have chosen not to examine this aspect of standards, focussing instead only upon their origination. Time and resources precluded any other choice.

Finally, the literature on the origination of standards – as opposed to the enforcement of standards – is sparse. This is particularly true if one wishes to include all the various types of standard setting, including voluntary, industry, regulatory and consensus standards. We cannot assume that our readers know very much about the general phenomenon of standard setting, for we have discovered that even regulators are often unaware of the origin of the standards they adopt and use. This situation requires us to assume little in terms of background knowledge about standards, and to provide a general introduction and discussion of standard setting.

In resolving these three problems, we have created a "book within a book". Chapter two through chapter seven can be read as a separate study of standard setting. We have another purpose, of course. For us, standard setting is an illustration of mandated science. Thus, chapter one introduces our argument about mandated science in addition to providing background information on the study. The concluding chapter of the book, chapter eight, provides a more detailed analysis of mandated science, based specifically on the case studies.

## The Design of the Study

This book is based on research carried out over a two year period. The methodology was case study research, a conventional approach in social science. Case study research is capable of generating a highly particularistic picture of situations or phenomena. The use of more than one case study permits researchers to make some comparisons, and to suggest some general conclusions. Case study research cannot, however, yield statistically significant results. To compensate for the relative lack of certainty attached to its conclusions, case study research offers a richness of detail, and a highly complex view of what is being studied.

In social science, case study research is often used to gain an insider's perspective upon events and situations. This involves the researcher coming to terms with how a situation or event is perceived and understood by its participants. Interviews and participant-observation are used as techniques for this purpose, as well as for establishing less subjective aspects of the event or situation. Case study research depends for its rigor upon a systematic review of all available documentary materials produced within and concerning any aspect of the situation being studied, upon interviewing as many people as possible, upon interview techniques that encourage dialogue and the expression of points of view, and upon detailed field notes from participant-observation. Our study of mandated science has used all of these techniques.

The research included six discrete cases, and a number of related events, organizations and controversies. We chose two standard setting organizations for study and comparison, but also examined some aspects of many more organizations with which we came in contact by studying the two chosen organizations. We chose four different chemical controversies for study, and in the process became acquainted with a wide range of chemical controversies. Our purpose in choosing chemical controversies for study was a limited one. We wanted to determine the role that standards, and decisions about standards, had played in these controversies.

In order to identify the standard setting organizations of interest to us, we traced the evolution of the standards used in environmental regulation and occupational settings in Canada. For many standards, this origin is in a little known organization called ACGIH, the American Conference of Governmental Industrial Hygienists. Standards developed by ACGIH are adopted in virtually every jurisdiction in the Western world; many are adopted without alteration or further scrutiny. We chose ACGIH as our first case study, and we attended two of its conferences and several different committee meetings over a three year period. We analysed its literature and documentation for particular standards and interviewed as many of its key participants as possible. Interviews were conducted in Cincinnati, Washington, Los Angeles, New York, Ottawa, Toronto and Vancouver.

The setting of standards also occurs in the international arena. To study how international organizations function, we chose the example of pesticide standards and an organization called the Codex Committee on Pesticide Residues (CCPR). CCPR is made up of national representatives who debate the scientific and trade implications of setting maximum residue limits for particular pesticides.

We attended the 1985 CCPR annual meeting in The Hague, and interviewed national delegates, officials and members of other standard setting bodies also in attendance. We conducted interviews with national and industry delegates, officials and sub-committee members in The Netherlands, and in Geneva, Rome, Paris, Lyon (France), Washington and Ottawa. We analysed the annual reports and documents of the CCPR, and of other related United Nations committees. In the study of CCPR, we also examined the work of organizations doing similar work. For this purpose, we interviewed officials and analysed materials from the European Common Market, the OECD, The International Agency for Research on Cancer, The International Program for Chemical Safety, and, in the United

States, The Environmental Protection Agency and Food and Drug Administration.

A description of standard setting organizations tells us little about how standards are used in the case of particular chemicals. To examine this problem, we chose four chemicals, and studied controversies concerning their standards and regulation. We encountered all the various public and interest groups, the institutional matrices within which decisions were made, the economic and legal implications of various proposals for regulation, and the political debates that emerged. We collected and analysed all of the scientific studies available to decision makers, including some that were never considered worthy of official attention. We read the transcripts of various hearings and court proceedings. We interviewed a significant proportion of the people involved in each of the controversies. As part of a separate sub-study, we interviewed more than eighty scientists in Canada who had been involved in the work of expert committees. We found that the history of decisions about standards was often buried in the story of the controversies that were recounted to us. We found this significant, and return to it later in a more general discussion about standards.

By good fortune, we had the opportunity to participate in two of the debates that we were studying. One of the members of the research team was appointed to a consultative committee, mandated to give advice to the government about the status of a particular pesticide. Detailed field notes were kept. As a result of observations about the consultative committee process, some further research – again of a mandated nature – was commissioned and a report prepared that recommended significant changes in the registration process for pesticides in Canada. The report was accepted in its entirety and has now been fully implemented.[15] Resulting from the research into pesticide registration, another member of the research group was given the opportunity to participate actively in the design of a risk management process for Canada.[16] Again, the onus was on the researcher to prepare the initial report, and again the report was accepted and is being implemented.

Exclusive of the separate study of scientists on expert committees, more than two hundred interviews were conducted for this study of mandated science. Almost all of the interviews were taped, and later transcribed. Material from the transcriptions is cited in this book. The interviews were conducted on a not-for-attribution basis, in order that interviewees would engage with us in a full and frank discussion. This procedure allowed us to get information that was not, strictly speaking, "on the record", and not simply the official positions of the organizations of interest to us. Information is provided in the text of this book about the position of the person being quoted, but names are not used in order to preserve the anonymity of those interviewed.

All chemical controversies occurred in Canada, but this is not a book about the Canadian regulatory process. There are two reasons. First, the chemicals under consideration are often produced outside Canada and only imported or formulated locally. The companies involved are seldom local in origin. Participants in the local debate are often drawn from other countries as a result. Decisions about standards taken by several American organizations and in the international arena have as much importance for understanding the controversies over particular chemicals in Canada as decisions taken in Canada. There is value in a study of the Canadian regulatory process; this book has quite a different purpose. It is written to be accessible to people with little knowledge about Canada or about the specific regulatory institutions in different countries.

Second, the focus of the research has been on the broader relationships among science, values and public policy. In spite of their differences from Canada, other countries experience many of the same problems as does Canada. If mandated science exists as a separate complex reality, as we argue it does, it exists in every jurisdiction, regardless of the differences in regulatory structures. The Canadian material is illustrative of that reality and of the problems encountered by regulators everywhere who depend upon science as a guide for public policy. The specific character of the Canadian political process – which has some elements in common with the governmental process in most industrial countries – is irrelevant unless one were to claim, as no one does, that the Canadian situation is unique. The chemicals we have chosen were also controversial in other countries.

Canadian material will be presented, then, because to do otherwise is to rob the study of its empirical content. The research was mainly conducted in Canada. Hopefully, readers from other countries will be enlightened by the similarities and differences posed by the Canadian examples without feeling the need for a crash course in Canadian politics.

## Specific Methodological Decisions

The chemical controversies chosen for study included those concerning captan, pentachlorophenol, lead and fluoride. We chose the pesticide captan for study because one of us had been appointed to a federal consultative committee for its assessment. The study of captan led us to CCPR, the international organization that sets standards for pesticide residues. We chose the controversy over pentachlorophenol in British Columbia, because pentachlorophenol is both a pesticide and an occupational hazard, and because a labour union played a significant role in creating this particular controversy about it. We chose fluoride because its standard in British Columbia was set by a workers' compensation board, but decisions concerning the fluoride standard were made by legislators. We chose lead, and the example of the Toronto lead controversy that occurred in the 1970's, because the lead case is unusual in the Canadian context. Normally, in Canada, chemical assessments are conducted by government departments or regulators. In the case of lead, lawyers were involved and the most significant assessment was one in which court-room tactics prevailed.

We were interested in these chemicals only in one jurisdiction and one time period. We have examined the broader controversies about some aspects of these chemicals only to the extent that the broader debates are relevant to the case studies that we are pursuing. For example, pentachlorophenol is linked to dioxins, and indirectly to Agent Orange and the chemical 2,4,5-T, but we have not examined the dioxin controversy nor the various assessments of Agent Orange, except as these assessments affected the controversy over pentachlorophenol in British Columbia.

Nonetheless, we should draw attention to the broader issues involved in the chemicals that we have chosen. Many scientific issues are common to all chemicals in our study. They are: the relevance of animal testing, the significance of testing for cancer causing potential using several different animal species, the problem of having conflicting scientific reports about a single chemical and the question of how to determine the cancer causing potential of any chemical. In the case of pentachlorophenol and fluoride, we face the additional question of the relative importance of environmental and human health hazards, of the role that safety precautions can play in mitigating the harm caused by a chemical in an industrial environment, and of the role played by workers and workers' compensation in standard setting.

We are not qualified to pass judgement on the toxic status of any of these chemicals nor to conduct a detailed assessment of the scientific literature with respect to these toxic substances. If we, as sociologists and philosophers, complain about the often naive assumptions made by interested but ill-informed commentators about social relations or values, we should be careful not to conduct a relatively ill-informed examination of the toxicological literature for the purpose of answering questions that even toxicologists cannot answer with confidence.

Nonetheless, we have surveyed the scientific literature and read the supporting expert testimony with respect to it. In two of the original case studies, we can take it as a given – the scientific opinion is overwhelmingly of one view – that the chemicals do pose significant dangers to human health. Few dispute the toxicity or effects of lead on human health; the issue here concerns the sources and techniques for its measurement and the measures appropriate for regulatory action. Few also dispute that pentachlorophenol causes environmental damage and poses some danger to humans exposed to it; the issue here is about the relative toxicity of the chemical and its normal impurities, and about the level of danger and the possibilities for its mitigation through preventative measures.

We believe, then, that these two chemicals, lead and pentachlorophenol, are legitimately under regulatory surveillance and that the public concern with respect to their toxic status has a basis in science. Beyond this very superficial report, we will say nothing about their dangers to either the environment or people. We are, however, qualified to examine the debate about standards. If we examine it as a debate, we can identify moral and other issues, and assess aspects of the science that are part of the debate about standards. These are questions that sociologists, communication scholars and philosophers can confront in any context, and certainly with respect to chemical standards.

## The Organization of the Book

In this book, we describe only two of the case studies of chemical controversies, the ones concerning lead and pentachlorophenol. We have done so for reasons of simplicity and length. A report on our study of captan is published elsewhere.[17] We have not yet published the case study of fluoride, although it is our intention to do so. It has been important for us, as researchers, to keep in mind the fact that our interest in these chemicals and the controversies surrounding them is as students of standards and of the mandated science used in standard setting.

We want to understand *how* the relationships between science, policy and social values are created in standard setting as an example of mandated science. We need to understand standards in terms of their use in trade, economic relations and in determining the relationship between the public and private sectors in the economy. We want to understand the role played by science and scientists in standard setting and, indeed, more generally in mandated science.

The organization of the book reflects this agenda. In the first section, we introduce the reader to some basic information about standards, and about the debates in which standards play a critical role. This basic information is necessary, because we have found much confusion about definitions in the debates about standards. In the second section of the book, we provide two case studies dealing with the work of standard setting organizations. In section three, we provide two case studies of controversial chemicals, in order to locate how decisions about standards were made in each. Each of the case studies is written so that it can be used as an independent unit, as a story in its own right.

In section four, we discuss standard setting in more general terms, focussing first upon the political dimensions of standard setting and then upon the way in which scientific information is incorporated – or not – in standards. We conclude the book by returning to the concept of mandated science that we outlined in chapter one. Using the four case studies of standard setting, we draw some tentative conclusions about the features of mandated science and discuss their implications for science, for decisions about values and for public policies.

Chapter Two

# An Introduction to Standards

There are standards for almost every product and activity. Standards exist for airborne contaminants, such as lead, and for contaminants in the drinking water, such as dioxins. There are standards for noise, in this case indicating the ratio of noise to silence within an acoustic environment. Standards are set for pesticide residues, so that only an approved amount of pesticide remains on a crop when it is sold or eaten. All of these standards deal with contaminants or pollutants, or with substances that are considered to be contaminants if they exceed a particular level.

There are other kinds of standards. The metric system is a standard for measurement. Apples are graded according to standards that classify apples in terms of size and appearance. There are standards for the purity of water and of chemicals used in manufacturing, standards for the use of engineers, standards for agricultural, industrial and home appliances, standards for the techniques that are used in sampling airborne contaminants, and standards for agricultural and industrial processes.

With such variety in standards, arriving at a single definition of the term "standard" is difficult. Environmental standards are numbers that indicate a designated ratio of a contaminant to the surrounding environment. Standards of measurement or those used in grading products are simply numerical points of reference for decisions about economic activity. These numbers indicate a norm, from which performance can be gauged. Standards for industrial activities are also norms, but they do not take the form of numbers. Instead, these standards describe an activity in such a manner as to indicate acceptable and unacceptable deviations from the normal practice of it. Perhaps this is the best way to define the general phenomenon of standards: all standards establish a norm but also the range of acceptable and unacceptable deviations from it.

Obviously there is great variation in standards. For example, some standards appear to be scientific in nature, and to be determined on a scientific basis. Other standards are created without reference to scientific information. Standards are developed by many different organizations, with many different goals and objectives. Indeed, sometimes more than one organization will set standards for the same product or activity. Social values are always reflected in a standard, but only sometimes are these social values related to health, safety or the environment. Standards can be voluntary or be identical to legal regulations.

## The Features of Standards

Because of the great variation in standards, it is important to identify their common features. First, all standards involve measurements, even if they do not take the form of numbers. Because of this fact, it is generally assumed that some form of technical or scientific determination is involved in setting standards. The very phenomenon of measurement ("standards") seems to imply that such a technical determination has taken place. All standards have the appearance of being scientific or technical in nature.

In pollution or noise control for example, it is assumed that scientists have been involved in deciding how much harm is created, and, in turn, how much harm can be tolerated with relative ease or safety. Countries with a registration process evaluate pesticides to determine their potential for harm, and this evaluation is seen to draw upon scientific information. In the case of a standard for electric kettles, a decision must be made about how hot the kettle can become before fire is caused. A technical determination is necessary before the *possibility* of growing apples of a certain size and appearance can be considered.

Only sometimes is the perception of standards scientific or technical based in fact, as our study of standard setting will demonstrate. This perception has important consequences, however. Even when little consideration has been given to either scientific or technical questions in the development of a standard, those who use standards – governments, companies and members of the general public – assume that a scientific or technical assessment has occurred. This is what we mean when we state that standards are the "scientific" component of the regulatory process, and when we assert that regulators often believe that a risk or scientific assessment has already been done when they adopt or alter a standard.

Second, all standards constitute points of comparison. An apple is compared with other apples and graded according to a standard. A pesticide standard is made more or less stringent in relation to an original level. Risks are compared, and these risk comparisons are used to originate or change a standard. The observation that standards reflect points of comparison is an important one. A common misperception is that adherence to a standard is a guarantee of public safety or of excellence. Assuming that the standards have been developed on the basis of an adequate scientific assessment of safety, adherence to a standard simply means that the product or activity involved is more likely to be safe than the product or activity that fails to meet the standard. In the case of apples, adherence to the standard "choice" means that the apples are more likely to be of excellent quality than apples that fail to meet the standard.

The actual comparisons used in standard setting are based on judgements about the acceptability of a product or activity (in comparison with a norm), rather than its excellence or desirability. The taste of an apple could be much less than either excellent or desirable, and still it might meet a prevailing standard for "choice" apples. An industrial process might well be dangerous, and still its use would meet the standard. The risks of a pesticide are compared with the option of not registering it for use. The pesticide might be registered, its pesticide residue might meet the prevailing standard, and exposure to the pesticide still constitute some risk to human health.

The comparisons are even more complex because what is being compared is not so much the apples, chemicals or activities in question, but average performances. For example, the speed at which a kettle heats up can be measured against a standard, in this case a measurement reflecting an acceptable average level of performance for kettles. But, some variation in the performance of kettles is inevitable; some kettles will heat more quickly than others, and the temperature that individual kettles eventually reach will differ.

The lead standard will reflect the amount of lead in human blood considered to be tolerable from the perspective of health, but blood lead levels in the same individual differ over time, and the blood lead levels differ among various individuals exposed to the same hazard. The lead standard is an average acceptable blood lead level, and this average acceptable level is compared with an average level of lead in the blood of an individual person or in a sample taken from a specific group.

In each of these comparisons, then, it is assumed that any individual kettle, or individual's blood lead level might deviate from the norm. These differences in individual performance are taken for granted. In deciding whether a kettle, or a particular blood lead level is acceptable, the judgement really concerns how much deviation from the average should be ignored, and how much deviation from the average is acceptable.

Thus, it is mistaken to assume that a standard constitutes a fixed point (a threshold) above (or below) which harm will always be avoided. It is also mistaken to believe that an individual product or activity conforms to a standard, simply because a standard exists and the products have been judged acceptable by it. The actual size, shape or nature of the activities that conform to a standard might well differ from instance to instance, and the effect of a pesticide in a specific case might well be different from that which has been approved. Standards take the form of specific measurements, but these measurements often reflect differences in performance.

The third feature of all standards is their close interconnection with economic activity. If foreign substances in the local creek are to be considered as pollution, there must be a polluting activity and an agent. If standards have been developed, the agent is likely to be a company and the standard refers to some industrial activity. If the pollution is to be cleaned up, this too will involve economic, and indeed often industrial activity. Noise pollution can be distinguished from loud noise, then, and standards generally exist only for noise pollution.

The kettle and the apple are products that are manufactured or grown commercially to meet specifications – standards – that have been determined in advance. Standards do not apply to apples from the tree in someone's backyard. In many cases, a product cannot be sold unless it meets a standard. In other cases, price differences will reflect the standards. The "choice" apple will sell for more than the "utility" one. Standards are measurements generated for the purposes of economic activity, and they govern contaminants, products, activities or industrial processes that are created or used by industry.

As a result, decisions about standards are always imbued with non-scientific judgements. Consideration of the economic consequences of a standard is intrinsic to its creation. As well, industry plays an important role in standard setting, even when standards for health and safety are being considered, because standards are designed with industrial activity in mind and are implemented by industry. Indeed, without industry participation – voluntary or coerced – standards would have little significance.

The fourth feature of all standards follows from the others. All standards reflect social values. In most of the examples chosen above, the standard is designed to protect health, safety or the environment. Nonetheless, a judgement is also made about an acceptable level of performance and about the economic consequences of imposing the standard. Standards also reflect concerns about the economic viability of an industry, and they protect the ease of trade. And because legal issues are often involved, and some standards are set through public discussion and government policy, questions about equity and fairness often arise.

Standards and social values are linked in a quite different way. We have just suggested that standards and excellence are different, but the popularly-held equation of standards and excellence can be dismissed too easily. It is correct to assert that standards set minimum levels of tolerable performance, and that such minimums only rarely meet a criterion of excellence. Nonetheless, the popular use of the term "standard" to mean a high standard does have an effect in the real world of standard setting.

The point is best made in reference of speed limits (which are standards, of course). The unwritten logic of speed limits can be seen in a fictional conversation with the driver who wishes to get to a destination quickly or explore the limits of the magnificent machine he (or she) has just purchased. It is as if such a driver is being told, "this fast, but no faster, because you jeopardize the safety of others by doing otherwise".

Many situations exist in which the speed limits are unreasonably high, and an accurate assessment of the safest speed along each stretch of roadway or for each type of car (assuming it is in good repair) is impossible. What starts out as a maximum speed limit quickly becomes a goal to be achieved, even in the minds of law enforcement officers. And in the mind of the speeding driver, it becomes a measure of obligation and of good behavior. In other words, once adopted, a speed limit, like any other standard, becomes a measure of *desirable (or excellent)* performance, a value to be sought after, a maximum rather than a minimum level of acceptable activity.

In sum, standards are norms for economic activity that are used for the purposes of comparison and decision making. All standards have several characteristics in common: they involve measurement and are perceived to be scientific or technical in nature. They involve two different kinds of non-scientific judgement, even before they are adopted by governments or regulators – a judgement about acceptability and a judgement about the economic consequences of adopting particular standards. They reflect an average level of performance for a class of contaminants, products or activities. Thus, standards provide norms for performance but also for acceptable deviations from it. Finally, standards are closely linked with social values, including values concerning trade as well as health and the environment. They are not measures of excellence, but standards become associated in the public mind with excellence once they are used.

To see how standards differ, it will be useful to have some specific examples in mind. In this book, we examine two standard setting organizations in great detail, ACGIH and CCPR. ACGIH sets standards for airborne contaminants in occupational settings. Its standards are called "TLVs", and they incorporate some scientific assessment. CCPR sets standards for pesticide residues on foods moving in international trade. Its standards are called "MRLs", and they are also developed with some scientific assessment beforehand.

Another standards organization is the International Standards Organization (ISO). The ISO is made up of delegates from national standards councils in many countries. The national standards councils generally are responsible for promoting standards and for the certification of standard setting organizations within a particular country. These organizations generally have some government support, but they are not usually government organizations, and private industry is often involved in them. The standards councils have a mandate to facilitate trade, particularly export trade. Thus, governments have a strong incentive to provide financial support for national standards councils and for the ISO, even where the national standards councils act independently of government.

ISO sets a variety of standards, mainly for the sizing of products and for the chemical composition of the chemicals used in manufacturing. ISO is also a proponent of the metric system. ISO arrives at its standards by seeking consensus among the delegates from national standards councils, and ISO only issues a standard if consensus can be reached. Scientific assessment plays a minimal role in ISO standards. ISO standards are guidelines, not regulations. Individual countries decide whether they wish to adopt ISO standards as national standards. Even if ISO standards are not accepted by all countries, their existence promotes compatibility in products and manufacturing, and thus supports international trade.

The Canadian Standards Association (CSA) is another typical standards organization. CSA is a private voluntary organization, established originally by engineers. CSA maintains testing laboratories, and issues a seal of approval for products conforming to CSA standards that are manufactured by companies under contract with the organization. CSA sets standards for products, but also for such activities as risk assessment.

CSA standards are consensus standards. The standard setting committees in CSA include representatives of four groups: the major producers, the major users, labour and government. A CSA standard will not be issued unless all four of these groups agree. Consumer representatives sit on advisory panels, and provide background information for the CSA standard setting committees, but consumers are not normally represented in the consensus that establishes a CSA standard. CSA standards are widely used by regulatory authorities, who review the standards – often in a cursory manner – and then adopt CSA standards as regulations. CSA standards are not themselves regulations.

When these four organizations are examined, it is obvious that standards are not the same as regulations. Some standards are regulatory, and some are simply guidelines. Moreover, some standards are considered to be guidelines until they are adopted by government as regulatory standards. For example, CSA standards for house wiring are adopted by Ontario Hydro, a publicly owned utility that provides most of the power to homes in Ontario and that acts in a regulatory capacity with respect to electrical installations. Hydro acts as the regulator, using CSA standards in most instances as the basis for its regulations and their enforcement.

Second, standards are created differently. CSA is a private organization. ACGIH is a private organization of government industrial hygienists. ISO is made up of representatives of national standards councils, who are themselves not necessarily government officials. CCPR is made up of national delegates. In the United States, the Occupational Health and Safety Administration (OSHA) and the Environmental Protection Agency (EPA) are both empowered to create standards, as well as to adopt them for the purposes of regulation. In Canada, government departments, not regulatory agencies, set or adopt standards as regulations.

Third, standards differ with respect to the level of scientific or technical consideration that enters into their creation. OSHA, EPA and the Canadian government departments take pride in the quality of their scientific assessment. CSA maintains a testing laboratory, but not all CSA standards reflect a technical or laboratory-based assessment. In CSA, achieving consensus among the interested parties is more important in the development of a standard than CSA's technical assessment. ISO standards are neither scientific nor based on a technical assessment.

Finally, standards differ in the extent to which they are implemented and enforced. Standards are often just guidelines, and even regulatory standards are not always enforced. Some regulatory standards apply only to some segments of an industry (in some cases, just to government organizations or to activities financed by government grants). Often more than one standard will exist for a particular product or activity. In this case, companies have considerable discretion about adopting any particular standard or about whether to enter a contractual agreement with a standards organization or not. Standardization or uniformity in performance as a result of standards is more of a fiction, strategy or proposal for action than it is a reality.

## Confusions in Terminology

The term "standard" has many different meanings, depending upon how and where it is used. Used in the United States, for example, standards tend to be equated with regulations. Pesticide standards can mean either the formal registration permits for specific pesticides manufactured and sold under a brand name or the more generally applicable rules for the sale of pesticides.

The terminology is confusing because the practice of standard setting differs in different countries, but also because public debates centre on particular – some would say peculiar – uses of the different terms. Because we will deal with a number of different standards and standards organizations, it is important to define the common terms used in standard setting, and to recognize the implications of the different labels that are used for standards.

A company marketing a pesticide requires approval in some countries before the product is released for sale. This approval is called *registration.* If the registration is limited to a brand named product or a brand named active ingredient, then the registration standard will be called a "*product specific standard or registration*". There are also regulations for the amount of peanuts in peanut butter. These are *regulatory standards* . The peanut butter company does not require approval before the peanut butter is marketed. The peanut butter is not a registered product. But peanut butter is subject to a number of rules, set in advance, concerning its peanut content. The rules are the regulatory standards.

The difference between registration and regulatory standards is that registration standards (or product specific standards) deal with specific products before they are allowed to enter the market. Products subject to registration standards must have permits before they can be sold or distributed. In contrast, a regulatory standard is simply a generally applicable rule.

A standard that is neither a registration nor a regulatory standard and that has no force of law is sometimes called a *guideline.* Other terms used for the same purpose are "objective", "code" or "criterion". Unfortunately, for our purposes, guidelines are also sometimes referred to as "standards" (as opposed to regulations, in this case) or as "voluntary standards." To complicate matters further, codes can be adopted as legal regulations by governments or as "regulations" by voluntary or by professional associations. This point is important because in the current debates about standards, the term "code" is sometimes used very differently, to distinguish between regulations ("standards") and self-regulation ("codes").

Even taking account the different uses of the terms "standard" and "code", there is an important distinction between them. In general, standards apply to products or activities. Codes tend to refer to behavior. For example, a standard exists for the materials used by engineers, and for the operations of a boiler they install in a factory. Codes exist for the conduct of the engineers as professionals. Thus, "codes" are associated with self-regulation, because the regulation of professional behavior is usually (but not necessarily) done by voluntary or professional associations.

The terms regulatory and registration standards apply to the implementation and enforcement of standards. There is another component to the standard setting process – the process by which the numbers or rules are generated initially. As we have already seen with CSA, ISO, ACGIH and CCPR, not all standards – even all registration or regulatory standards – are generated in the same manner. Some are set by government departments, some by governmental agencies, some by private voluntary organizations (and later adopted as regulations), and some are created by organizations that have government participation, but which are not themselves governmental bodies.

Three kinds of standards can be distinguished on the basis of the type of organization that originates them. *Voluntary* and *consensus* standards are standards originated by standard setting organizations. Voluntary and consensus standards are different from each other because of the manner in which they are set. Voluntary standards are developed by a single organization. Consensus standards are developed through some form of consensus and compromise between competing interest groups. Neither voluntary nor consensus standards are the subject of government regulation when they are developed, except inasmuch as administrative rules may govern what will qualify as a consensus standard for the purposes of enacting regulations. Both can be adopted later by government as regulations. Finally, some standards are originated by government organizations, rather than by consensus or voluntary bodies. For the purpose of contrasting them with voluntary or consensus standards, they should probably be called *governmental* standards.

In the literature and in everyday practice, governmental standards are often referred to as registration or regulatory standards. This is confusing, because voluntary or consensus standards can be, and often are adopted for use by regulatory agencies or government departments. The TLVs, for example, are developed by a voluntary organization, but they have been adopted by a number of government agencies and departments as regulatory standards. CSA standards are consensus standards that are adopted for use as regulations by public utilities such as Ontario Hydro. It is even more confusing because standards developed by government departments or agencies can be viewed as voluntary, and used as guidelines not regulations for conduct. Thus, in this book, the terms "voluntary", "consensus" and "governmental" standards refer only to the process by which the standards are originated, not to the eventual status of the standard as a regulation or as a guideline.

Finally, it is necessary to distinguish between the terms *standardization* and *harmonization*, since both terms are often used in the international debate about standards. Standardization refers to the acceptance of an identical level of performance (one or more standards) by a number of different groups or countries. It is not always possible, or desirable, to achieve standardization. Harmonization refers to the process of making different standards compatible – although not always identical – with each other for the purpose of trade. For example, different countries might maintain different registration standards for pesticides. Harmonization can be said to be accomplished in this situation when these different standards no longer constitute a barrier to their trade. Quite often harmonization involves some degree of standardization.

We suspect our readers will be relieved to hear that the purpose of distinguishing between these types of standards is a limited one. In the literature on standards, all these different terms are used. They are seldom explained, because it is generally assumed that the readers share the same knowledge about the political systems of the countries in which the standards are being developed as the author does. We will use only the terms registration, regulatory standard, and guideline to refer to the way standards are implemented. We will use the terms voluntary standards, consensus standards and governmental standards to refer to how standards are originated.

When *we* want to speak about the differences between registration and regulatory standards, we will make our intentions clear. When we want to speak about the origination of standards, and distinguish it from the implementation and enforcement of standards, we will say so. When we feel that it is important to distinguish between voluntary and consensus standards, and to view them as clearly distinct, we will be specific in the text. Usually the term standard will refer to the general phenomenon, and not to any particular type of standard.

## The Data Problem in Standard Setting

If standards are to be used to protect the public, or even to deal with problems in trade, it is important that the information used to develop them is adequate. Unfortunately, such is not always the case. Some countries now require manufacturers to file information with a government agency on product chemistry. These records are often not very extensive; and few resources are dedicated for monitoring the data, and for requesting further assessment. In any case, for most industrial chemicals, governments do not have a mandate to demand toxicological testing, even if the chemical or product in question is subject to regulatory standards.

Countries that register pesticides require extensive information before registration is granted. The reality of the registration process is, however, that most pesticides now in use were registered before extensive assessments were required, and they were registered on the basis of scanty information. Analytical techniques were not developed until recently to the point where many hazards could be identified. Some of the data problems in registering pesticides also stem from the fact that the data produced for the purposes of registration is proprietary, the private property of the manufacturers which commission the studies and submit them for the purposes of registration. Companies seldom make their studies public, except when they submit studies to the scrutiny of a public hearing.

The companies who own the data protect its confidentiality for economic reasons. The cost of generating the scientific research required for registering a single product can exceed ten million dollars. If this research were made freely available, the companies argue, then competitors could simply use the information to register their products – the companies call this "me too registrations". Registration of a product is similar to patent protection, in that it limits the competition to those whose products have been approved. As a result, companies are not likely to want their studies published in the open scientific literature or released to the public.

These problems with data can all be attributed to the lack of information at the time when chemicals are being reviewed, problems in the registration process caused by the proprietarial nature of the data or lack of resources for adequate scientific assessment. They are what is commonly referred to as "the data problem" in chemical assessment and standard setting. The problems with obtaining adequate scientific information for decision making, and the lack of public scrutiny of such information, undermine public confidence in standards.

There are other kinds of data problems in standard setting. One is concerned with the trustworthiness of the data. In the press, attention is often drawn to the case of the Industrial Bio-test (IBT) Laboratories, which was discovered in 1977 to have falsified research being used to support the registration of a number of pesticides. We cannot know how common such faulty testing might be, in spite of assurances by companies and regulators that it occurs rarely. Codes of good laboratory practice, and on site inspections in the United States and a few other countries, were instituted after 1977 as a result of the IBT scandal, and they represent an improvement in public safety.

The public concerns about the validity of scientific tests cannot be dismissed, however, if companies continue to include IBT studies in their submissions of scientific evidence about their products, long after they have been discredited – as has recently occurred in a Canadian federal inquiry into the pesticide alachlor.[1] The story of the asbestos standard shows that companies sometimes will conceal their knowledge about the dangers of specific chemicals, regardless of the legal and financial consequences of doing so.[2]

Another data problem stems from the nature of the research itself. Scientific studies generally indicate – at best – whether a chemical has potential to cause problems in laboratory animals. The animal data is used to gauge human harm. The real question in the public mind is whether problems actually emerge in the home or workplace, and whether public health will suffer either immediately or at any point in the future. This latter kind of information – namely, that concerning the human consequences of chemical hazards – is much harder to obtain and to evaluate.

One approach to answering the questions about human harm is to conduct epidemiological studies. In such studies, evidence is sought to see if a correlation can be established between exposure to a particular chemical and health problems. The best example of the use of epidemiological research for the purposes of public policy is in connection with cigarette smoking. As the tobacco industry is quick to point out, studies that illustrate a positive correlation between cigarette smoking and cancer do not prove that a causal link exists. Only research on what happens when cigarette smoke is metabolized within the body can provide evidence that cigarettes "cause cancer", and such research requires more information on the biological origins of cancer than is yet available. In the interim, epidemiological studies provide evidence, but they do not establish cause.

The problem for epidemiologists of establishing cause exists because the human exposure to cigarettes occurs in natural, not laboratory situations. In a natural setting, the experimental conditions cannot be easily controlled. It occurs because the effects and their measurement occur some time later – often much later – than the exposure. As well, epidemiological data are often drawn from incomplete medical records, personal recall, or personal descriptions of symptoms. Subjective factors compromise the findings of epidemiological research, even if the studies have been conducted with rigorous attention to conventional scientific procedures. Because of these problems, a great number of epidemiological studies are usually required to establish persuasive scientific evidence about a causal link between exposure to a particular chemical and harm. A single epidemiological study is seldom sufficient in itself to persuade policy makers or those involved in standard setting that a standard should be revised or a chemical restricted.

In spite of these difficulties, epidemiological studies are very useful, because they often provide the only available direct evidence to link human exposure to chemicals to the harm such chemicals might cause. Epidemiological research is also very persuasive, because it deals directly with people, not laboratory animals. But epidemiological research cannot provide assurance that all problems in the home or workplace have been identified, at least without a great deal more time and resources than have, as yet, been available to conduct it.

Insurance and company medical records – when they exist – can provide another indication of the health problems of workers and some evidence about their exposure to particular chemicals. The number of insurers and companies is great, and their record keeping practices are varied. Medical and insurance company records are private, and no regulatory agency in any western industrial country has the power to compel systematic disclosure of all the personal information such records contain. Records of court cases about injuries or health problems are more likely to provide evidence about the practices of taking legal action rather than the incidence of workplace dangers.

The data problems confound the attempt to conduct scientific assessments credibly. When resources are lacking for detailed study of all potential chemical hazards, regulators will often fall back upon the public's identification of problems as the best guide for action, reasoning that this method is as adequate as any other for determining when a chemical should be reviewed. If, however, public knowledge about how to identify problems is scanty, if procedures for notifying government authorities about problems are lacking, or if each individual considers his or her problem to be unique, then problems will go unreported. Regulators often state that no problems have been reported about a particular chemical in the workplace. Quite often, we simply do not know whether such problems exist, because the regulator in question has no way of identifying them.

Even if they have the best intentions, those who set standards are hampered by problems we have already described: the lack of resources, the proprietary nature of the data and the very considerable costs of protecting the environment and workers. They lack the resources or legal powers to develop adequate epidemiological data, and are often wary of conclusions drawn from epidemiological research. Standards are often opposed by workers, who see in them an acceptance of contamination of the workplace, or a threat to their jobs. Those who set standards function in a world characterized by conflict and by insufficient information. The standards they originate reflect this fact.

## The Debates about Standards

There are three current public debates about standards. Although this book does not focus on any of these debates directly, it is important to identify what is at issue in them, and the relationship between our research on standards and the arguments being advanced in each case.

In one debate, issues in standard setting are presented in terms of a conflict between regulation and deregulation. Regulatory standards, it is argued, constitute a particularly onerous form of government intervention.[3] Regulatory standards set rules for industry that are unresponsive to the dictates of a freely operating market system. Their implementation brings a host of record keepers and inspectors to industrial worksites, most of whose activities can simply be seen as bureaucratic red tape. If regulation were more effective, the argument goes, then some justification might exist for this government intervention and bureaucratic red tape. Because regulation is inefficient and ineffective, regulatory standards should be replaced by other methods of achieving public health and safety.

One proposal being advanced in this debate about standards is to replace regulatory standards with guidelines set by voluntary or consensus bodies, or with codes of good practice.[4] It is argued that voluntary standards are enforceable, while regulatory standards are not, because the people who originate the standards voluntarily would not set standards they were unwilling to comply with. It is suggested that the public interest could also be served if such standards were set by a consensus body, if all the various interest groups had representation on it. Codes of good practice, it is also argued, are industry-imposed forms of policing. Industry has an incentive to abide by these codes, either because public good will depends upon it or because codes can easily be replaced by regulatory standards if they are not adequate or because access to government subsidies or support is dependent upon compliance.

We will return to the debate about the value of regulatory standards later. When we do, we will argue that there is a substratum, another level, to this debate that is not immediately apparent in the press or academic literature. More is at stake, we will argue, than a philosophical commitment to free markets or government regulation. Indeed, neither of these alternatives, regulation or deregulation, seems to us, on the basis of our research, to have much relevance in the actual creation and implementation of standards.

The second debate is about prescriptive and performance standards. Even if one were to accept that regulatory standards – set and enforced by government – were beneficial, the question would still remain as to whether such standards should be prescriptive or performance standards. Prescriptive standards in occupational health and safety and environmental regulation are set in advance of the activities to be controlled by the standards. Examples of prescriptive standards are the rules set for achieving clean air and clean water. Sometimes these prescriptive standards are called "emission" standards. Other examples of prescriptive standards are the rules concerning the height of smokestacks from a factory and the rules for the equipment to abate pollution.

Performance standards in occupational health and safety and environmental regulation deal with the actual exposure of the public to a particular substance and, sometimes, with the harm resulting from that exposure. Examples of performance standards are the rules concerning the amount of contaminants in the air caused by emissions from a factory. These standards are sometimes called "impingement" standards, because they focus on the contaminant at the point of its impingement on the environment or people. Performance standards are also set to determine the degree of injury to workers that will be sufficient to permit them to claim workers' compensation payment or insurance.

The debate about prescriptive versus performance standards – simply put – is about whether to use a standard for the smokestack or for the harm resulting from the contamination emitted from it. With prescriptive standards, there are easily accessible mechanisms for enforcement. With a minimal allocation of resources, it is easy to see when violations of the rules have occurred. The rules can be set far enough in advance so that the construction of the smokestack cannot proceed unless a company abides by them. The implementation of prescriptive standards requires extensive record keeping, however, and an active government inspectorate. Prescriptive standards are synonymous with government intervention.

Performance standards pose as an alternative to prescriptive standards, because they deal with contaminants only if harm is caused. Performance standards require far less record keeping. Yet, performance standards are much more difficult to use. It is difficult to measure the level of contaminants in a worker's lungs, and even more difficult to determine how much of the contaminant is absorbed from the lungs into the body or how much harm is caused. Even if accurate measurements could be obtained, individual workers are susceptible to injury at different levels of contamination. What injures one worker has no effect upon another. Often the effects of exposure to a chemical will not be evident for many years.

As much as we seek to minimize harm and to use whichever standards are best suitable to achieve that objective, it is apparent that use of neither prescriptive nor performance standards provides a simple solution to the problem of how to achieve these objectives. It may be that the choice is not between either prescriptive or performance standards, but between a series of options for assigning the responsibility and allocating costs of injury caused by chemical contaminants. The debate about prescriptive and performance standards is an important one, and we will return to this debate in the closing chapters of this book.

The third debate concerns the enforcement of standards. Here again, the focus of concern is the level of government involvement with standards. Different proposals have been made – and some implemented – about ways to enforce standards without requiring government inspectors or without resort to legal mechanisms of enforcement. One such proposal involves placing a price on units of pollution, and taxing pollutors according to the number of units of pollution they produce. Another allows companies to market their allotment of permissable units of pollution, so that a company that succeeds in reducing its pollution can benefit economically by the sale of its pollution rights. Both these proposals substitute economic incentives for regulation to ensure the enforcement of standards.

The issues raised in the third debate, about the enforcement of standards, are important ones, for several governments have embarked on a search for nonregulatory means of ensuring compliance with governmental standards. This study deals with the origination of standards only, so the third debate will not be discussed at any length in this book.

## Standard Setting as an Example of Mandated Science

The features of standards that we identified originally – that they measure acceptability, that they are connected to economic activity (and, as we will see, legal issues as well), that they reflect social values, and that they have the appearance of being scientific – make standard setting an ideal laboratory for the study of mandated science. Standard setting often involves the incorporation of science or science-like information. Standards are seen to be scientific or technical in nature. Yet standard setting involves value choices and policy decisions. It takes place in a number of different organizations that are neither scientific laboratories nor exclusively devoted to the making of public policy. These organizations have a mandate, which is to produce the measurements that can be used for guidance and for regulatory purposes by industry and government.

We will confine our discussion primarily to the standards used for pollutants. In arriving at these standards, it is (or should be) necessary to decide how dangerous a particular pollutant is. Once the danger has been estimated, it is then necessary to determine how much of the substance can be tolerated by people exposed to it. How much danger should be tolerated? Assuming some people are affected more seriously than others, how many and which segments of the population will attract most serious concern, and, indeed, what level of caution is advisable in matters of public health and safety?

Answering these questions will be difficult, because any health and safety standard imposes costs that will eventually be passed on to consumers. If the costs of abating pollution are too high – the standard is too stringent – industry insists that factories will be closed. If the standards are set too low, we are likely to be accused of being captive of the major industrial interests whose pollution had attracted our attention in the first place. Moreover, if we are convinced that a health risk exists, is it appropriate to consider costs at all? Should we protect human health and safety from serious risks at any cost?

Assuming a standard is to be set – and some level of risk is to be considered acceptable – who should do it? Should we make the standard as a regulation and enforce it vigorously? What about the costs of stringent regulations or the easily demonstrated difficulties of vigorous enforcement, given that any single infringement of a regulation is unlikely, in itself, to create significant health hazards?

In theory, we can take each of these questions in turn. We can set standards by first conducting a scientific assessment, then by engaging in a public debate about values, and, finally, by designing a method for their implementation. The actual practice of standard setting is neither so rational nor so easily segmented.

The relationship between science, values and public policy in standard setting is considerably more complex than the relationships we have just described. The only way to illuminate the actual character of these relationships is to examine how standard setting occurs.

The next four chapters are case studies of standard setting organizations and of chemical controversies in which standards play an important role. Each case study tells its own story, and these chapters are designed to present it rather than to examine the more general phenomena of standard setting and mandated science. It is in the the last two chapters that we draw conclusions from all of the the case study research and examine the complexity of the relationships in mandated science quite specifically.

Chapter Three

# In the Eye of the Storm
# Case Study One: The American Conference of Governmental Industrial Hygienists

*with Edwin Levy*

Canada has originated only a few standards. Neither provincial nor federal authorities have the resources to develop a full range of standards, either for environmental or for occupational contaminants. Canadian regulators usually adopt standards developed elsewhere, or rely upon the existence of such standards as an alternative to regulation and for workers' compensation awards. In this, Canada is typical of all countries. Pressing for an answer about the origin of most of their standards, we discovered a little known organization called the American Conference of Governmental Industrial Hygienists (ACGIH).

ACGIH is an American voluntary organization of industrial hygienists and academics. Industry personnel, people associated with the insurance industry and military personnel also attend ACGIH meetings, and sometimes take part in ACGIH deliberations about standards. ACGIH is associated with the professional group of industrial hygienists, the American Industrial Hygiene Association, but does not itself represent or regulate a profession.

The objectives of ACGIH are to promote sound industrial hygiene practices, and to promote workers' health and safety. ACGIH sets standards for many workplace hazards, including for airborne contaminants in the workplace. The standards for airborne contaminants are called "Threshold Limit Values" or "TLVs". The TLVs are designed to deal with airborne and occupational hazards but TLVs have been used as the basis for environmental standards, including standards for contaminants in the soil. The terms "Threshold Limit Value" and "TLV" are both copyrighted.

ACGIH standards are used by many regulatory authorities in Canada and elsewhere, including the Occupational Health and Safety Administration (OSHA) in the United States. In such cases, ACGIH standards – which are guidelines and not regulatory standards – are often adopted as regulations and enforced as such. ACGIH standards are also used as guidelines by industry. Although ACGIH has no public profile, the TLVs are the standards most often referred to in the popular press. A press report might say that the "accepted standard is..." , the same number as the TLV. ACGIH standards are used extensively throughout the western world.

At first we were not impressed by what we heard about ACGIH. Its deliberations do not resemble the open, and frankly adversarial relations of regulatory standard setting that many have come to expect. The fact that few knew much about ACGIH was disturbing, for it suggested that the public interest might be compromised by the invisibility of ACGIH's activities. We decided to investigate for ourselves the organization that has such an important role to play in setting standards.

There is no question that the scientific information used by ACGIH to make decisions about industrial pollutants was often limited. The active involvement of industry in setting ACGIH standards for the pollutants was easy to document. Nonetheless, in the past – and even today – such organizations as ACGIH fill a vacuum between public knowledge and public concern. Organizations such as ACGIH act when government regulators cannot or will not.

ACGIH has been secure in its assumption that it is not held legally responsible for its standards. This voluntary nature of ACGIH gives it a particular kind of power. It is the power to identify where concern should be focussed, and to inject some practical measure of safety in the workplace. Remove ACGIH's standards and, historically at least, the workplace would have been considerably less safe. To put the assessment of ACGIH and its standards in context, a brief account of the history of the organization and its early debates is necessary.

## The Early History

ACGIH was formed in 1938, and began its standard setting activities a few years later. The members addressed the problem of contaminants first by developing guidelines or codes (of conduct) for practicing industrial hygienists. By 1942, these codes were seen to be problematic, because they placed responsibility for ensuring workplace safety on the hygienist as a professional, rather than on the originator of the contaminants – industry or its management. ACGIH codes were then supplemented and later replaced by standards, which designated limits of tolerance for particular chemicals (and physical agents) in the workplace.

The ACGIH initiative was not the first of its kind. The American Standards Association was already developing limits for exposure to chemicals in the workplace, as were the US Public Health Service, and an informal group of hygienists in Boston. Nor were the ACGIH standards unique, since a survey had already been conducted of State health units in the United States to develop an initial list of standards, and a second list of standards was furnished to ACGIH by the U.S. Public Health Service. Indeed, ACGIH saw itself as an alternative to other standard setting bodies whose "machinery at best grinds out its grist finely and slowly."[1]

Early ACGIH publications stressed the difference between the ACGIH, governmental agencies and industry-based standard setting organizations. For example, ACGIH rejected the idea of setting regulatory standards, its members emphasizing the limits of government action. As well, as ACGIH noted, regulatory standards are often derived from animal data. Human experience with contaminants is discounted as insufficiently scientific. ACGIH standards, on the other hand, were to be based on the practical experience with contaminants and on easily observed – but not necessarily scientifically documented – problems with human health as a result of them. As the Director of the Bureau of Industrial Hygiene in Detroit and one of the first members of the ACGIH standards committee, noted

> All of us doing field work know that if samples are to be taken for any contaminant in a plant, we must produce a limit, right or wrong, for the consideration of management. Otherwise they feel quite rightly that we were wasting our time and theirs taking samples in the first place.[2]

ACGIH also rejected the consensus approach of the industry-based American Standards Association, arguing that health and manufacturing standards should be set differently. Health standards, ACGIH argued, should be set by a committee of experts, while manufacturing standards might properly reflect – as they did in the ASA – a consensus of manufacturers. ACGIH membership was also different from that of the ASA. ACGIH members were government industrial hygienists. ACGIH argued that its members were free from the conflicts of interest as a consequence.

Nonetheless, the issue of the influence of industry in ACGIH was raised in the early meetings of ACGIH. A member of the TLV Committee of ACGIH provides us with an example of ACGIH's concern about industry participation. A new and more stringent standard for carbon tetrachloride had been proposed in ACGIH by an advisory committee of experts. It was circulated to the manufacturers for their comments. In light of a very negative response from industry, it was withdrawn. One participant in the expert committee is quoted as saying:

> Certain individuals who were members of our Massachusetts advisory committee and also of the ASA (American Standards Association) committee reversed their belief, previously expressed, and voted for the 100 parts per million. Again, I am told that one of these individuals did so with the statement that it was done more or less under protest and for the sake of securing unanimity.[3]

Later, the issue of industry participation was raised with respect to the membership on the TLV committee, and as a resolution to the problem, industry participants were designated as consultants and were not permitted to vote in ACGIH or on specific standards.

In the early period of ACGIH's history, the informal nature of the ACGIH standard setting process was seen to be its greatest strength. By using its annual meetings to discuss and adopt standards, ACGIH felt that it could keep standards responsive to new information. This was an optimistic claim, but a reasonable one in light of the difficulties experienced by government regulators in responding to new information. By publishing a revised yearly list, it could also discourage the adoption of its standards as "legal codes" or government regulations. Once established as legal codes, ACGIH felt that its standards could not be altered easily. It was an ironic concern, given that the later success of ACGIH rested upon the value of its standards for regulators seeking a quick method of developing regulations.

## The Active Phase

The most active period in ACGIH was between 1961 and 1968. In 1961, 280 TLVs had been published. By 1970, the number of chemicals for which standards were set had risen to 500. (In 1981, there were 599 standards for different chemicals, and by 1984, 615). Many of the revisions of standards also occurred in the 1960s. As well, less than 2500 TLV booklets had been distributed yearly before 1960, and by 1970 that number had risen to 30,000.

Between 1961 and 1968, ACGIH acted in a regulatory vacuum in the United States, in spite of the existence of the governmental Bureau of Occupational Safety and Health (BOSH). BOSH had the necessary jurisdiction, but its activities were hardly impressive. As one person from ACGIH noted:

> BOSH did not have a legal mandate to go into industry. A lot of times they did this through the backdoor, through state programs. Also they did so with the agreement that there would be no disclosure. You could not do that today. BOSH had very little budget. It wasn't until the mid-1960s that it reached a million. In 1968, the budget was still (made up of) 88% personnel benefits. The organization really couldn't accomplish much except for the fact that it had dedicated people at various levels.[4]

BOSH officials relied heavily on ACGIH:

> Mainly they (BOSH) accomplished a great deal through their professional associations, ACGIH and AIHA (American Industrial Hygienists Association). (Their officials) didn't have money but (they) had time. You couldn't find ACGIH on an organizational chart of the US government, but per person, I think it had more impact than most others. I'd stack it up against any other organization.

In practice, there was relatively little division of labour between ACGIH and BOSH. Standards were passed back and forth between ACGIH and government agencies for consideration and adoption.

During the 1960s, scanty documentation was published to support ACGIH standards, and manufacturers exerted considerable influence upon ACGIH. The report of the Committee on Threshold Limit Values in 1964 tells the story:

> The full Committee held two, 2 day meetings in Washington D.C., December 2 and 3rd, 1963 and March 2 and 3, 1964. Following the first meeting, a "Notice of Intent" was publicized by AIHA in their Newsletter and by ACGIH in their Bulletin Board, and subsequently by the Manufacturing Chemists' Association, in an effort to inform hygienists in industry and government of the proposed changes in the Threshold Limits for 1964. The response by industrial representatives to the "Notice of Intent" at first drew mainly inquiries of the reasons for the suggested changes; later replies, numbering about two dozen, took issue with certain of the proposals. By and large, these replies were not accompanied by substantiating data. The Committee, however, was pleased to receive a number of suggested additions to the Threshold Limits list with supporting data adequate to develop recommendations.[5]

The documentation to support ACGIH standards was initially contained in a small booklet, compared with the 500 page volume of documentation issued by ACGIH today. "Personal communication" was more important than documentation. Almost all of the documentation of the standards set in the period from 1961 through 1968 requires updating, but this process is by no means complete yet, nor is it being undertaken on a systematic basis.

Two other key issues for ACGIH were first raised in these active years. Should ACGIH be a professional association, and should its standard setting activity be co-ordinated with AIHA, the existing professional association? The issue of ACGIH's status as a separate professional body was not resolved, and still arises from time to time today. At the ACGIH conference in 1964, however, a proposal was received from AIHA that the organizations join in setting standards. In rejecting the proposal for joint action, ACGIH stated that its mandate was: "to recommend from time to time to the Conference threshold limits of toxic air-borne materials as a consensus of the levels at which exposed persons will be protected from adverse health effects." It was noted that AIHA, as a professional association, was bound by consensus of its members, while ACGIH saw itself as relying heavily upon expertise in the development of standards. Today AIHA has its own standard setting committees and standards, suggesting that the conflict between the two bodies with respect to their standard setting activities is not yet resolved in spite of co-operation between the two groups.

## The Transition Period

The creation of two new government agencies for occupational health and safety (OSHA and NIOSH) in the United States in the late 1960s threw ACGIH into a quandary. These governmental bodies seemed likely to replace ACGIH, yet ACGIH was reluctant to withdraw from its standard setting activities. An ACGIH member, who was also on staff at one of the government agencies, describes the situation in 1968, as follows:

> When the OSHAct was passed, the leadership in NIOSH realized that some changes would have to be made.... I think NIOSH tended to look towards OSHA (rather than ACGIH) as its closest liaison. NIOSH was instituted under the same Act as OSHA and they had a kind of legal working relationship. But by and large you had a different cast of characters developing numbers, and when you get individuals who feel their missions are different, you get some competitiveness, some jealousy.

ACGIH's response was to distinguish itself from the two governmental agencies. In 1968, ACGIH was using the NIOSH address; it immediately acquired its own. The activities of ACGIH were then moved out of what was then NIOSH (formerly BOSH) facilities and a process of disengagement was begun. The situation was described by the same informant, as follows:

> At the time of OSHA's creation, there was a lot of soul searching at ACGIH. We wondered whether we should just fold up our tent and go home. There was a lot of encouragement in that direction coming from NIOSH. NIOSH felt that now it had legal responsibility for establishing criteria for standards, that ACGIH's TLV committee had done its job well, but that now we were in a new era and NIOSH superceded us. There were a lot of people at NIOSH who felt that way and weren't afraid to express it to the TLV committee and ACGIH itself. I was on the Board of Directors, but I think even more discussion was taking place in the TLV committees. It ended up with a wait and see attitude for a couple of years. By the mid-1970s, there was a realization that the new system was not going to be responsive to current problems.

NIOSH was supposed to produce criteria documents – documents assessing the scientific information that would be used to support standards. OSHA was supposed to use NIOSH information to develop appropriate regulatory standards. In fact, the situation was not very different from that which ACGIH deplored initially. NIOSH was slow; OSHA was slower and even more dependent upon a political will to proceed. There was still a need for voluntary standards that could be put into place quickly, and be revised easily. NIOSH recommendations, OSHA standards and ACGIH's TLVs were often identical. ACGIH felt justified in claiming that its standards were as adequate as those of the governmental agencies. As one member noted:

> One of the things that became evident was that NIOSH was publishing criteria documents, but OSHA was not promulgating standards. So the recommendations in the criteria documents were more like guidelines. They were enforced somewhat under OSHA's general duty clause. Subsequently, they fell apart. In the 10–12 year history of NIOSH, they have about 100–150 chemicals (for which there are standards). There are 600 TLVs. ACGIH can generate a number and implement it much more readily than government can. Every time you try to change a standard at the federal level, the lawyers are in there. You know about the races to the courthouse.

Because the 1968 TLVs had been incorporated into the Walsh-Healy Act, an act concerned with, among other things, some workplaces under federal jurisdiction, a great many ACGIH standards were adopted by OSHA. Even today, these standards form the backbone of OSHA regulations. Our ACGIH informant describes the situation, as follows:

> I don't think it was accidental. There had been several attempts over the preceding years to promulgate an OSHAct...and it was just a question of time as to when there would be a national occupational health and safety program. The language of the OSHAct specifically provided for the Secretary of Labor to promulgate as interim or start-up standards, national consensus standards, that had already been promulgated under certain Acts including the Walsh-Healy Act. Now the people in the Bureau of Labor Standards who were responsible for promulgating those standards were the same people who were going to be responsible under OSHA for setting the interim standards. Many of these people were ACGIH members but that doesn't make it an ACGIH decision. These people knew what was coming down the road and that they would have a job to do. If you had that responsibility, what would you use?

There was a hidden issue in deciding whether OSHA would adopt ACGIH standards. Technically speaking, ACGIH standards are not consensus standards, but the legislation establishing OSHA required that only consensus standards be adopted. As one informant suggested:

> Section 5(A) of the OSHAct mandates the Secretary of Labor to adopt, without dealing with title 5 of the Administrative Procedures Act, as soon as practicable, any of the consensus standards already established in federal regulations...Some argue that the Secretary had discussions (before adopting the standards). Others argue that the adoption was automatic because the big employers were already using these standards.

There was some discussion in ACGIH about whether to adopt a consensus method, but ACGIH did not do so. As one person described the situation:

> Stokinger saw the legislation (OSHAct) required consensus standards from that point on (for the purpose of their being adopted as OSHA regulations). So he looked around and appointed industry and union representatives on the TLV committee for the first time. I don't think that this is appreciated. Stokinger was wrong, but he thought he could make the TLV committee (into) a consensus body if there were industry and union representatives.

ACGIH standards were not consensus standards, but they were adopted by OSHA regardless. The incorporation by OSHA of the TLVs was meant to be a temporary measure, a means of ensuring the existence of regulatory standards until a proper regulatory assessment and process of their promulgation could be undertaken. As OSHA standards, however, the TLVs become embedded in a process that had limited capacity for change. Many OSHA standards today are still the ACGIH standards from 1968. They have not been revised, although several of the ACGIH standards have been. As OSHA standards, as well, the TLVs lose their apparent connection to ACGIH, with the result that ACGIH's influence is easily underestimated with respect to standards in effect today.

In the mid-1970s, ACGIH's standard setting activities revolved around a new kind of standard, the STEL or short-term exposure limit. The TLVs were originally designed to deal with exposure to chemicals over the length of the working week. They were inadequate to protect workers from short-term (for example, fifteen minute) exposure to the same chemicals. The question whether to set limits for short-term exposures (STELs) was first raised in 1965, with a comment in the Conference transcript:

> The recent adoption of short-term limits for the air of industry by the Commonwealth of Pennsylvania into a regulation brings into focus the need for the Executive Committee of the ACGIH to give consideration to the extension of the present activities of the TLV committee to include short-term limits, in order to provide a centralized agency for the origination of such limits.[6]

By 1975, ACGIH was ready to act on these short-term exposure limits, and act they did. In one year, STELs were introduced for 84% of TLVs we surveyed. In 1977, 28% of those STELs were withdrawn. The question of whether to issue STELs was and is highly controversial.

## ACGIH Today

Today, ACGIH is a vibrant, if small and little known association. It describes itself as "an organization devoted to the development of administrative and technical aspects of worker health protection" and emphasizes its contribution to the "official industrial health services to industry and labor".[7] Its membership of approximately 2500 consists of professional personnel in government agencies or educational institutions engaged in occupational health and safety programs. Its objectives are to promote sound industrial hygiene practices, co-ordinate industrial hygiene activities through federal, state, local and territorial industrial hygiene agencies, to encourage the free exchange of experiences and ideas, to collect and make available information and data to government industrial hygienists, and to hold annual meetings.

The organization maintains a small office in Cincinnati, mainly for the Executive Secretary and the clerical staff necessary to support the publication program. Most of the work is conducted through the Board and its committees and their subcommittees. The work of the TLV committee on chemical substances in the workplace is best known, but there is a TLV committee for physical agents (such as noise, vibration etc), and industrial ventilation, and committees for agricultural practices, sampling and monitoring, engineering control technology and, recently, a committee on law and ethics. Each committee can have several subcommittees. The committees report annually at the ACGIH conference that is held in conjunction with the conference of AIHA. In recent years, a few hundred people attend these annual meetings, which are open to the public, from an AIHA conference population that now exceeds 5000. A transcript of the annual meeting is published. The TLV committees meet during the year between the annual meetings, but other ACGIH committees meet only as required.

The budget of ACGIH gives a good picture of the organization. Literature production, publication, printing and postage account for half of an annual budget of approximately one million dollars. A further $220,000 goes into salaries, most of which are for staff to produce and disseminate the publications. The most important ACGIH publication is the standards or TLV booklet. Four-fifths of the TLV booklets are sold in the US, mainly to industry. Nonetheless, the TLV booklets are available worldwide and more than 100,000 are sold annually. In addition to the TLV booklet, a series of specialized publications and reports from special symposia are sold.

The travel expenditures of the committee members – their only remuneration – account for about $130,000. Literature searches, on the other hand, were budgeted for less than $5000, indicating that ACGIH committee members draw upon their own resources to gather and analyse scientific materials. Revenue for ACGIH comes mainly from publication sales ($886,000 projected in 1985), and membership dues. ACGIH declared a small profit (2.2% of gross income) in 1984.[8]

The 1984 report of the TLV committee provided us with a second picture of ACGIH's activities today. It included a notice of seven proposed standards for new chemicals and thirteen revisions.[9] A large number of STELs were deleted, and fourteen standards, previously discussed, were added to the list of adopted standards. Thirty-nine chemical standards were listed as under study. Typical issues that were discussed included the measurement of asbestos and the appropriateness of ceiling limits. Comments were solicited to assist the committee in its deliberations and in the development of draft documentation. It was noted that draft documentation is used by the TLV committee to determine what action, if any, to recommend on a given substance or question.

## Membership of the TLV Committee

Since decisions about standards are made by the TLV committees, their membership is of great interest. Only a total of fifty-seven people served on the TLV chemical contaminants committee between 1961 and 1983-4.[10] Committee members held tenure for long periods of time, as Table 1 indicates.

Information is available on all but seven of these individuals with respect to their formal qualification. Thirteen of the members have been Ph.D.s; fourteen have been MDs; three have been professional engineers and only five hold advanced degrees in industrial hygiene or public health. Thirteen members list

TABLE 1
Length of Tenure
Chemical Substances TLV Committee

| Years on Committee | Appointed between 1961–68 | Appointed between 1969–76* | Appointed between 1977–84** |
|---|---|---|---|
| 0–3 | 3 | 4 | 6 |
| 4–7 | 5 | 2 | 11 |
| 8–11 | 3 | 8 | 1 |
| 12–15 | 2 | 3 | – |
| 16–19 | 4 | – | – |
| 20–23 | 4 | – | – |

* Of these 3 were still members of the TLV committee in 1983–4.
** 13 were still members.

some formal association with industry, five with the US military and fourteen with governmental regulatory authorities.

These statistics can be rounded out with a more descriptive account of TLV committee membership:

> Now we have several OSHA people on the committee ... Mr. A, who is the latest chairman is a professor of veterinary medicine ... and he came out of the Air Force. He's got a doctorate in veterinary medicine and he worked in toxicology for the airforce..until his retirement ... and now he's on the faculty at a U.S. university. Mr. B is an MD who is in occupational medicine ... Mr. C used to work for Exxon Chemical, I think, and he retired, so he's now teaching ... and he's on the committee. So effectively he's from academia. Anyone from academia and government is not considered to be from industry, you see. Mr. D is a toxicologist and he is NIOSH. An excellent man and a good researcher. Mr. E is an industrial hygienist in the Navy ... a practicing industrial hygienist like myself. Mr. F just retired from (a) state department of health. Mr. G is an MD and doctor of science ... I think he's with the veterinary school up there. Mr. H came out of the airforce, and he was with environmental engineering. Ms. I is in the OSHA office ... Mr. J is with OSHA. Mr. K has a doctorate in public health. He's with NCI and functions somewhat as an epidemiologist ... (etc.)

Personal contact seems to be the most important method of committee selection. One member of the committee described the selection process, as follows:

> Selection is a kind of haphazard process. Generally, I think someone says: do you want to consider him for the committee? So it's by recommendation, more or less. The whole committee does not consider candidates; the chairman does.

Another member told us how he was selected:

> I joined ACGIH in the '50s and I went to the hygiene meetings; I didn't go to the medical meetings (he is an M.D.). I was more interested in the environment at the workplace and control of the work environment. There are too many physicians who take that for granted. They are more interested in clinical matters, in how to diagnose

> aplastic anemia.... I opened my mouth at a few meetings and, as a result, got put on committees and eventually got elected Chairman of ACGIH.

This picture of informal selection of ACGIH members was borne out by our experience as researchers. As academics, we too were welcomed to committee meetings, and had the occasion arose, it was not inconceivable that we would have been invited to serve on an ACGIH committee.

When asked about the role of industry members, a former member of the TLV committee commented:

> Yes, they are treated in the same way (as other members). They would be asked to make input into documentation. They, as professionals, comment freely about whether they like or dislike something. But when it comes to the crunch, when the committee is going to decide, "Are we going to classify this as a carcinogen? Are we going to set the level at 0.1 or at 1?" They don't have a vote in that.

Industry members are involved in ACGIH as consultants to its committees, in initiating standards or their revision, in providing most of the data upon which standards are based, in preparing proposals and draft documentation and in making representations before the committees. Industry members do not vote in ACGIH, but their influence is obviously significant. Another former member saw the issue of industry participation in ACGIH somewhat differently:

> Sometimes on the TLV committee you might go along with something you are not too keen about, but in the long run you might achieve your objectives. For example, I wanted to reduce the TLV for (chemical X) from 200 to 100 ppm a number of years ago and I got shot down by the Committee because one fellow who was a consultant (and is now deceased) made a flippant comment that industry could not live with 100 ppm. Yet they were living with 100 ppm for (chemical Y) and all other solvents that are closely related.
>
> I try to set standards on the basis of physiological and toxicological activity, not on whether industry can live with it, but it took a number of years before I was able to accomplish that reduction.... You have to be cognizant of the needs of the employer too. To be very frank about it, if you knock industry out of the picture, we would all go back to living in caves and tents. So you have to be realistic and practical in this business. There's no such thing as being a purist. How pure does a society want its air? How much are they willing to spend for it? It all comes down to economics, doesn't it?

At least until recently, labour has been a reluctant participant in ACGIH. In the early 1970s, one member of the TLV committee was a formally designated representative of labour, and he acted in the capacity of liaison for four years. A TLV committee member who was present in the early 1970s described labour's involvement at the time:

> Mr. X of the autoworkers was a member. He was a pretty level-headed guy. I think he contributed meaningfully to the TLV committee. Clearly both union and industry people were far outweighed by the government people. Then organized labour began to criticize the TLVs and they felt they were in an awkward position, since they had members on the committee. So Mr. X retired.
>
> The TLV Committee chairman got in touch with another well known hygienist from organized labour and asked him to join. He never really said no, but he never joined either. It's just their way of boycotting us.

At the meetings we attended, there was no sense that labour participation was being actively encouraged, although few committee members were drawn from unions. The absence of an active labour presence in ACGIH is both striking and noteworthy.

ACGIH committee members are often governmental regulators but surprisingly, this situation has also been problematic. For some time, it was felt that regulatory officials should not serve officially as ACGIH committee members, however unusual this might seem. It was argued that ACGIH members who were also government regulators had a dual, and potentially conflicting mandate – from government and from a private voluntary organization. For a short period, regulators were forbidden to serve as committee members and asked to serve as consultants, in the same manner as industry participants. This suggests that ACGIH and government regulatory bodies are to some extent in competition. The concern about regulators acting as committee members has subsided, but greater involvement from academics is now being encouraged.

Control over committee membership is exerted through the chairmen (or women), positions appointed by the ACGIH Board. Membership is voluntary; only travel expenses are covered. ACGIH itself has no facilities for conducting research or scientific evaluation. ACGIH is dependent upon those with resources from other sources to carry out its work. Committees are as expert as their members make them and have only those resources as are made available through their members.

## Standard Setting in ACGIH

Generally, the initiative for setting a new or revised standard will come from industry. One former TLV committee member said:

> Virtually every – not all, but virtually every – substance for which a new TLV has been developed has been at the request of the industry in question, of the manufacturer or producer – in most cases Dow, Du Pont and the like. Very seldom would the chairman or the secretary of ACGIH suggest a chemical. (A consultant) would come in and say his company is making a new pesticide or whatever...They obviously had data from the company's perspective – which was in their view good data. At least data we could not have many questions about. We could have questions about the conclusions, but not the data itself.

A committee member described the process for initiating standards as follows:

> The TLV committee still subscribes to the idea that if we have a problem with a chemical we need a number for it. So let's put it on the list of intended changes and see what information and criticisms come forth. If that number is going to pinch profits from a company we will hear from them. They know that if OSHA adopts this value they will have to get down to it. And they may have medical and other records that support a higher number. It's been a constant plea for 30-40 years to get big companies to co-operate and provide information about their products. One way to get information is to propose a number and let them respond.

ACGIH has neither a formal process for setting priorities, nor formal criteria for developing standards. A committee member described how decisions are made:

> Committee members get lists of chemicals coming in and people will say, "where did this one come from?" Someone else will say, "Joe Schmoe from this chemical company brought the chemical to our attention and they've been having some problems; here's what they have been doing to try to protect their workers and so forth." It's being used throughout the country as an intermediate or straight chemical, and we'll look at it.
>
> It can't be a single purpose manufactured chemical. In other words, it can't be a chemical used by one company only. But if it's a chemical in widespread use – we're talking about thousands of pounds of stuff – and we get some feedback that there's some problems in the handling of the chemical, we will look at it. Frequently a company will come in and say we want you to establish a TLV for such and such a chemical. So we find out they are the only people making it and they are using it for internal consumption. We won't do that; we won't establish a TLV.

Another TLV committee member stated:

> There were informal priorities. There always has been a tremendous amount of communication between the chairperson and people in the field, industry and sometimes labor. There were lots of letters and phone calls that went back and forth.

Once a decision is made to establish a standard, committee members, including their consultants from industry, seek further information in support of a proposed standard or its revision. Preparing the draft documentation can be assigned to any member or members, including the consultants. Preparing the draft documentation often includes reading the relevant literature and consulting those deemed to be most knowledgeable about the chemical. In many cases, we were told, the industry consultants were the best prepared to assemble the documentation package and to propose a TLV. Certainly industry is always consulted and industry's presumed or actual reactions to proposals play a role in ACGIH deliberations.

All of the recommendations for new standards and revisions are discussed and voted on first in the committees and then at the annual meeting. Based on our observations, it would be a mistake to call the discussions at the annual meeting thorough. In general, the annual meetings accept the recommendations proposed by the committees. Nonetheless, questions and concerns are raised from the floor and, occasionally, a matter is sent back from the annual meeting to a committee for further study.[11]

## ACGIH Standards

ACGIH sets a number of different standards for chemical contaminants. The TLV-TWA, or *Threshold Limit Value-Time Weighted Average,* is a measure related to the concentration of an airborne chemical for a "normal 8 hour workday and a 40 hour workweek, to which all workers may be repeatedly exposed, day after day without adverse effect".[12]

The TLV-STEL, or *Short Term Exposure Limit* is "the concentration to which workers can be exposed continuously for a short period of time without suffering from 1) irritation, 2) chronic or irreversible tissue damage, or 3) narcosis of sufficient degree to increase the likelihood of accidental injury, impair self-rescue or materially reduce work efficiency, and provided that the daily TLV time weighted average is not exceeded." It supplements the TWA threshold limit values where there are acute effects from a substance whose toxic effects are primarily of a chronic nature. STELs are recommended only where toxic effects have been reported from high short-term exposures in either humans or animals. and are defined as "a 15 minute time-weighted average exposure which should not be exceeded at any time during a work day even if the eight hour time-weighted average is within the TLV."[13] In discussing STELs, ACGIH notes that such exposures should not occur more than four times a day and there should be at least 60 minutes between exposures.

There are some chemical contaminants for which the effects cannot be offset by periods of low exposure. ACGIH sets a *"Ceiling"* ("C") for these substances. For TLVs without such a ceiling, time weighted averages permit some *"excursions"* above the limit. That is, TWAs are based on exposure over a time period, averaging the high and low peaks of exposure. If high peaks in exposure are likely to cause health problems, the designation of a "C" and an excursion limit are used.

ACGIH maintains that there is not enough toxicological data (even in the government regulatory assessments) to derive very specific excursion limits. Its approach is "intuitive", an attempt to ensure that excursions be held "within some reasonable limits". A mathematical formula can be used to derive excursion limits from TLVs, in order to ensure that they are related to "the variability generally observed in actual industrial processes." ACGIH uses a simplified version of this mathematical formula. And where toxicological data are available to set a short-term limit, a STEL is used.[14]

ACGIH has expressed interest in a new measure, *Biological Limit Values*. Biological Limit Values are limits directly related to the human harm caused by chemical contaminants, as opposed to limits for the presence of airborne contaminants. The TLV booklet also provides information on the tests that are available to detect individuals who are hypersusceptible to a variety of industrial chemicals.[15]

## Controversies about Standards

Any standard setting organization dealing with chemicals must deal with two controversial issues: whether to set standards for carcinogens, and how often to revise its standards. The question of how to handle carcinogens is a difficult one for ACGIH because many scientists and regulators believe that threshold limits cannot be developed for chemicals with carcinogenic potential. Their argument is that a very small exposure to a carcinogenic agent can act as a triggering or promoting factor. Very low levels of exposure to these chemicals can result in the development of tumors. Indeed, some scientists question whether a threshold can be set for any chemical contaminant that results in human harm.

ACGIH accepts the concept of a threshold, of course, and ACGIH does set threshold levels for potential carcinogens. Rather than identify the potency of a carcinogen, ACGIH uses different categories to indicate the carcinogenic potential of specific chemicals. For example, they distinguish between substances of high and intermediate carcinogenic potential (depending upon the evidence of eliciting cancer by three or more, or less than three animal species at specific dose levels) and low carcinogenic potential (showing evidence of causing cancer in one animal species by any one of three routes of administration under specific dosages and conditions).[16]

ACGIH takes pride in its capacity to respond quickly to the need for standards and their revision. To what extent is this pride justified? We took a sample of ACGIH standards, by tracing the establishment and revision of every tenth standard in the 1984 TLV booklet. Table 3 charts the pattern of standards revision.

The total possible number of revisions – assuming all sampled chemicals were revised each year with respect to each standard – is 21,168. The number of revisions in our sample was 157. Revision of standards is infrequent and often slow, even when new research is available. The great bulk of ACGIH standards remain in force for many years without revision. Based on our observations, we think ACGIH is typical of all standard setting organizations – including regulatory bodies – in this regard. Without prompting or special circumstances, standard setting bodies revise their standards infrequently.

TABLE 2

Percentage of chemicals in sample for which standards were introduced

| | |
|---|---|
| 1961 | 47% |
| 1962–68 | 32% |
| 1969–75 | 10% |
| 1976–82 | 13% |

Percentages have been rounded.

TABLE 3

Number of revisions in standards (not including new standards)

| | |
|---|---|
| 1961 | 5 |
| 1964 | 22 |
| 1967 | 24 |
| 1970 | 9 |
| 1973 | 18 |
| 1976 | 38 (includes 15 STELs) |
| 1979 | 30 |
| 1982 | 11 |

## The Status of ACGIH Standards

We should begin with the description of the status of ACGIH standards that is contained in the TLV booklet and reiterated by every person from ACGIH we interviewed:

> The Threshold Limit Values, as issued by ACGIH, are recommendations and should be used as guidelines for good practices. Wherever these values (of whatever year) have been used or included by reference in Federal and/or State statutes and registers, the TLVs do have the force and effect of law.[17]

The TLV booklet makes it clear that there are differing levels of scientific assessment for differing chemicals. ACGIH says:

> These limits are intended for use in the practice of industrial hygiene and should be interpreted and applied only by a person trained in this discipline. They are not intended for use, or for modification or use, (1) as a relative index of hazard or toxicity, (2) in the evaluation or control of community air pollution nuisances, (3) in estimating the toxic potential of continuous, uninterrupted exposures or other extended work periods, (4) as proof or disproof of an existing disease or physical condition, or (5) for adoption in countries whose working conditions differ from those in the United States of America and where substances and processes differ. The TLV-TWA should be used as guides in the control of health hazards and should not be used as fine lines between safe and dangerous concentrations.[18]

ACGIH could not make itself any clearer about the limitations on its standards. If an industrial hygienist relies upon the TLVs, and if his or her judgement is questioned in a court of law, ACGIH argues that the organization can and should take no responsibility. The TLVs are guidelines; they must be interpreted by professionals in the field; professional judgement is necessarily involved; it is the professional's judgement that is properly addressed in a litigious context. A former TLV committee member described the accepted view of ACGIH standards:

> The TLV is only a guideline and it is only one guideline. Because of the variability of responses of individuals you will see some people affected at 10 ppm and others not until 100 ppm. The TLVs in the booklet are levels that are set to protect humans from acute effects. But this might not protect humans from adverse effects due to chronic exposure....
>
> If you were to take arsenic trioxide and blow it at certain concentrations on animals all day long, you would never get lung cancer in animals, yet arsenic trioxide is a known human carcinogen affecting the lungs. If you set a standard on the basis of animal data you would be all screwed up. If you go to smelters where workers are exposed to it, you will find cases of lung cancer. You have to be careful. It takes someone with insight into what these numbers mean.
>
> ...When I go into the workplace I look at a number of factors: how long has the worker been exposed, the age and general physical condition of the worker, the sex, and the colour of the worker, etc. These last two are relevant because of differential effects. For example, some chemicals affect light skinned people more than dark skinned; some the other way around. You don't learn these sorts of things at school. You gradually pick this up by being in the field, on the job.

Indeed, ACGIH – as an organization – has never appeared in a court to defend its standards. Legal advice has been sought, probably in response to the increasing possibility that industrial hygienists will be sued and, certainly, as other standards bodies have been implicated in legal cases. For now, little concern is expressed about the legal status of ACGIH standards, and little concern seems necessary.

There is another side to the story. It has its ironic elements. In the early days of ACGIH, it was much easier to maintain that ACGIH standards were simply guidelines, since ACGIH was quite explicit about the high degree of estimation ("rules of thumb") involved in their development. Today, ACGIH devotes considerable attention to the scientific basis of its standards, and – not surprisingly – defends its standards as being as solidly scientific as any others. The practical side of professional judgement, so critical to ACGIH's understanding of its standards, becomes less important to the extent that the scientific basis of the ACGIH standards is stressed. We shall return to this point later, for it is an important one for the discussion of mandated science.

## The Use of ACGIH Standards

Regardless of the provisons in the ACGIH booklet about the adoption of ACGIH standards, the fact is that they are often used outside of the United States. Although ACGIH keeps no record of this international distribution or the use of its TLVs, several meetings have been held in Europe with ACGIH and other standards bodies connected, for example, to the European Common Market. Widespread and international use of the TLV booklet has double-edged consequences for ACGIH. On one hand, it strengthens the credibility of the TLVs and ensures a healthy income from the sale of TLV booklets. On the other hand, automatic adoption of TLVs as regulatory standards is likely to change the legal status of the TLVs.

We found widespread evidence of the extensive use of ACGIH standards in Canada and elsewhere. In an interview with a Canadian workers' compensation board official, we asked how the Ontario compensation board determined the standards for its compensation decisions. We were told that the Ontario Ministry of Labour was responsible for standard setting, but that few official (i.e. Ministry of Labour) standards existed. "For the others?" The ACGIH booklet was pulled out from the desk drawer. Did this official know anything about ACGIH or the development and status of its standards? The answer was negative. "Were the limitations on the applicability of ACGIH standards addressed specifically?" It was evident that not much attention had been paid to them.

Ontario does have a process for standard setting, but very few substances have been "designated" for the development of a new standard. For the most part, Ontario accepts the OSHA standards, which are, in turn, mainly the same as ACGIH standards. As one Ontario Ministry of Labour official told us:

> Ontario has chosen to go the crazy OSHA route of creating standards substance by substance, adversary by adversary. It will take them years to do what the TLV committee has done. Instead they should adopt the "standards by reference". This means, for example, that the Ontario Health and Safety Act would say: "we will adopt the TLVs recommended by ACGIH except in those instances in which we specify a difference" That way you could adopt the latest edition instead of the 1933 edition staying in the books to 1968 (a reference to OSHA).

In B.C., the adoption of ACGIH standards is quite deliberate. An official from the compensation board described the situation, as follows:

> The TLVs are the starting point and the focal point for evaluating permissable airborne concentrations at BC WCB (Workers' Compensation Board). To do otherwise would mean setting up a complete, separate organization to evaluate every one of the chemicals in use. Theoretically that is what we would like to do. In our latest revision of our regulations, we attempted to take some of the substances that were either missing from the ACGIH values or where we did not agree with the TLVs and make up our own. One of these was gasoline and another was wood dust,

> especially cedar dust. Concerning the latter, we relied on some work by Dr. A. and our own evaluations. We decided to cut the ACGIH level in half. The reliance on ACGIH also explains why we do not have separate values for the different chlorophenols. When we next revise our regulations, we will look more closely at that.

When asked about the similarity between B.C. standards and those of ACGIH, he continued:

> If you take a look at our permissable concentration levels in Appendix A of the *Industrial Health and Safety Regulations* you will see that they are based on ACGIH. We did not use the term "TLV" because it is a registered trademark. If we had used the term, we would have had to include the preface, the preamble of the TLV booklet, and that says that the numbers are only guesses and that they are to be used only as guides by knowledgeable people. You can't put out a regulation and say, "well, it's just a guide". So that is why we had to say, "Look, we'll take their numbers, but we will make them our own." So without changing the numbers, except in the odd instance, we took them in.

We have been told that "the United Kingdom used to adopt the entire list of TLVs holus bolus, but now they say they are developing their own" and that the ILO (International Labour Organization) is aware of TLVs and uses them in their publications. We have no way to check who uses ACGIH standards. Neither does ACGIH. Nor can we trace direct links between specific regulations in various countries (or as used by United Nations agencies), because, as the B.C. case illustrates, the relationship between adoption of ACGIH TLVs and their appearance as regulations is complicated by the way in which the regulations are adopted and described.

We do assert, and ACGIH concurs, that ACGIH TLVs are in wide distribution. They are used, more than anyone is likely to document, at least as a starting point for standards throughout the western industrial world and probably in third world countries as well. We suspect that ACGIH standards are more than a starting point and that their true regulatory status is masked by the way in which they are generated and distributed by ACGIH.

## Discussion

### *ACGIH Standards and Science*

ACGIH uses measurements of airborne contaminants, measures of potential human exposure in other words. These measures will provide a different picture of contamination, depending on how and when they are taken. ACGIH cannot assume that monitoring of air samples is continuous or systematically pursued in all industry contexts. Or even most of them. ACGIH cannot assume that adequate monitoring equipment is available to the industrial hygienist on the job or that resources will be allocated by any specific company for detailed analysis of the samples. However much we might wish it otherwise, the situation in industry does not always, or even often (some claim), support good industrial hygiene, at least with respect to the airborne contaminants that are not governed by regulatory standards.

For these reasons, ACGIH must make decisions about the most practical and appropriate ways of gauging the levels for contaminants. The time-weighted averages can easily be criticised, because they can result in short-term exposures to very high levels of the contaminants. The STELs have been the subject of vigorous debate. The excursion limits are a product of some less than rigorous assumptions about the range of levels of contamination that can be sustained in a working day. The introduction of ceiling levels was equally controversial, even assuming that ceiling levels could be monitored and enforced. The assumptions are supported by some scientific data, but also by extrapolations at best and guesswork otherwise.

ACGIH does not present itself to the public as a scientific organization, even if it argues that ACGIH standards are as adequate scientifically as are most regulatory standards. ACGIH members are cognizant of the difference between toxicological assessments and those of ACGIH, which today include a toxicological assessment but also other kinds of scrutiny. For example, the choice of an 8 hour basis for the time-weighted standard is not a simple adjustment of a toxicologically-derived measure. It cannot be adjusted on the basis of toxicological data to reflect a longer or shorter working day. The TLV-TWA bears no direct relationship to a toxicological measure of risk. A TLV committee member provided some details of how use of the TLV-TWAs might cause problems:

> Suppose you are using a particular solvent (name given) that has a particular TLV of 100 ppm. But then you switch to another solvent, whose TLV is 350 ppm. Because of the TLVs it might seem like the new material is safer. But when you look at the physical properties of the materials you find out that because of a lower boiling point of the new solvent, it is actually more of a hazard in the workplace. So the TLV booklet should not be used as a cookbook. You have to go in there and look at the situation and analyse it from many points of view.

The TWA is based on a notion that *most* workers will not be affected in a negative manner, but the definition of most is problematic. A former committee member argues:

> The booklet states that the TLV's are levels below which a majority or the vast majority of the workers will be protected. This is not merely a guesstimate; this is absolute fantasy. We cannot tell whether 15% or 5% or 1% of the workers are protected at the TLV level. Even for substances that have long-term or chronic effects, the TLV is based not only on short-term effects, but on acute ones.

All ACGIH standards are based on operational definitions. They are as good as the operationalizations themselves. To the degree that the operationalization of the standard reflects the actual situation within the workplace (and with respect to the potential for harm), the standards are reasonably sound. Neither ACGIH, nor any practicing industrial hygienist, has much control over the variables that make the operationalizations reasonable, however. ACGIH must simply construct its best estimate of the circumstances in which the contamination might occur. It protects its estimate as a measure of ensuring workplace safety by saying that the companies and their industrial hygienists must exercise their own judgement and not rely merely upon the TLVs to determine whether the workplace is safe.

The most vigorous debates within ACGIH have been concerned with how to measure contamination, and not with the toxicological status of particular chemicals. This is in contrast to regulatory standards. ACGIH is primarily (not exclusively, of course) concerned with the experiential side of chemical exposures, while debates about regulatory standards often (though not always) are concerned with the dangers of the chemicals themselves. Reliance on experience as a basis for standards does not mean that ACGIH standards are less scientific than regulatory standards. If this were the case, it would indeed be surprising to discover that regulatory agencies and ACGIH often arrive at the same conclusions.

In assessing the scientific content of ACGIH standards, it was important to remind ourselves of some simple points. Scientific research is based on systematic observation. Problems arise in any scientific assessment – toxicological or other – when extraneous factors prevent the observer from being very systematic. In the case of all standards – ACGIH, regulatory standards and others – there are many factors that confound the process of observation. In the case of ACGIH, it is easy to document how such factors undermine a scientific assessment. The basic operation of using experiential data for assessments is not itself an unscientific one, however, whatever problems in measurement are introduced by the use of such data. Today ACGIH responds – we think rightly – to the criticism that its standards are not sufficiently scientific by drawing attention to the methodological problems that confound all assessments – including toxicological assessments – and by stressing how by using experiential data ACGIH can err on the side of caution in protecting workers' health and safety.

### *ACGIH Standards and Risk*

The first ACGIH standards were in the form of codes for practicing industrial hygienists. Such codes placed significant responsibility upon the professional industrial hygienist, as an individual, for creating the safe workplace. Codes can refer to the activities of the manufacturer, too. Again in this instance, codes deal with the behavior of the manufacturer as an individual. ACGIH codes were quickly supplemented, and then supplanted by standards called "maximum allowable concentrations" ("MACs"). Because the term "maximum allowable concentration" was thought to have inappropriate connotations with respect to encouraging contaminants, MACs were replaced by TLVs.

In setting TLVs, ACGIH took its emphasis away from the activities of hygientists as individuals, and concentrated instead on the quality of air in the workplace. The measurement involved in implementing a TLV was only indirectly concerned with the behavior of individuals, and had no immediate connection with the professional conduct of either the hygienist as a professional, or the manufacturer as an individual. The TLVs were designed to facilitate control – usually technical control – over contamination in the workplace. Industrial hygienists using TLVs deal with sampling and monitoring the air in the workplace and with equipment designed to control contaminants or exposure to them.

Today there is some interest in ACGIH in the development of biological limit values. These values relate to the acceptable levels of absorption of the airborne contaminants by the body. With the potential adoption of the biological limit value as the standard, a further change is occurring – or will occur if the biological limit value approach is adopted more fully – in the orientation of ACGIH standards. With the full adoption of biological limit values as standards, it is necessary to arrive at some estimate of how individual workers absorb chemical contaminants.

In such cases, it is often necessary to conduct tests that are intrusive of the worker and identify his or her propensities to chemical absorption. For example, blood tests can be used to determine how much contamination has been absorbed into the bloodstream from inhalation of a contaminant. Urine tests are now routinely required from employees in certain jobs. A comprehensive system of BLVs would combine the toxicological assessment of harm with an assessment of the absorption of chemicals by individual workers. The latter assessment raises significant questions of civil liberties – for example, the rights of workers to refuse blood and urine tests.

Biological limit values are only the first step in the development of quite a new and different kind of standard. Let us assume that it becomes possible, with new analytical techniques and further scientific experimentation, to link the presence of chemicals in an individual's body – chemicals that can be identified in terms of biological limit values – with specific diseases. This is the approach taken in biological and genetic screening. Once it is determined that some groups have a genetic or biological predisposition towards contracting diseases from the absorption of specific chemicals, it is possible to decide whether such groups of workers should be allowed to be exposed to the chemical contaminant.

With the use of biological and genetic screening, individual workers can be excluded from employment before specific damage from contaminants occurs. They can be excluded if they are members of a group with propensity to experience harm from contamination. Someone – a purposefully vague term – must decide whether groups at high risk should have the right to work in situations where they are at risk.

With any biological standard, a portion of the responsibility for implementing standards is shifted from industry to the individuals who are likely to be exposed. The measurement of contamination is no longer a technical one, but includes aspects of personal risk and freedom. Control over contamination becomes the responsibility not just of the particular industry producing the contaminants, but of the worker who might be predisposed to harm (and who must decide whether to work) and the employer (who decides whether to hire – or fire – a particular worker who is at risk). With biological and genetic screening, society must decide what to do with groups who are genetically at greater risk from exposure.

An example of the use of rudimentary biological standards is easily provided. In recent years, decisions have been made to prohibit pregnant women – or women of an age where pregnancy is possible – from certain employment opportunities because they are at greater risk than others from contaminants in those jobs. For the individual woman worker, opportunities for employment are curtailed whether or not she is or intends to become pregnant. Her status as a woman is as much of interest in setting a standard as the technical measurement of the contaminant in the workplace.

The choice of restricting certain jobs with respect to gender is a relatively easy one, for it is still widely accepted that society has an interest in the health of children. The ethical issue becomes clearer if it is determined that Orientals are unusually susceptible to a particular contaminant. In such a case, the standard used to determine employment would be based on race, and it is a standard that is applied to the individual potentially at risk rather than to the level of contaminant in the workplace.

ACGIH has not replaced the TLV with the biological limit value, although some of its members advocate doing so. Moreover, biological limit values are only a first step in the development of biological and genetic screening. The point we are making is not specifically concerned with biological standards or even with the illustrative history of the evolution of ACGIH standards. With each kind of standard, something different is being measured. In the case of codes, the measurement is in relation to individual conduct. In the case of the MACs and TLVs, the measurements are of the actual contaminants, and the measurement is largely a technical one. With Biological Limit Values, an attempt is made to measure the human harm, based on levels of exposure to contaminants. Sometimes this involves testing that is intrusive to the individual. With biological and genetic screening, the focus of concern is either the individual who is at risk, or a group of people who might be at risk because of their biological and genetic makeup.

TABLE 4

Standards and the measurement of risk

| Codes | MACs/TLV | Biological limit values | Biological and genetic screening |
|---|---|---|---|
| Oriented to the conduct of professionals and management | Oriented to the contaminants in the workplace | Oriented to measurements of human exposure and harm | Oriented to the individuals and groups at risk |

The relationship between the different types of standards and the assignment of risk can be illustrated graphically in Table 4.

The social and political implications – and legal consequences – inherent in the choice of different types of standards should now be evident. In the case of codes, it is the professional or the company manager who bears the brunt of the responsibility for safety. The TLVs focus on the contaminant and not the people involved in creating or measuring it. As time-weighted averages, however, the TLVs are not a precise measurement of either exposure or harm. Much effort of a highly systematic nature would be required to prove that a specific company (or individual) was responsible for human harm on the basis of their having permitted contaminants to exceed the TLV. From a legal standpoint, there is some question whether a legally supportable link could be made between exposure and harm by simply using the TLV and the measurements taken in conjunction with it.

The biological limit values focus concern upon the individuals who are harmed, but because the BLVs are currently extrapolated from TLVs, they too are not an accurate measurement of the connection between exposure and harm. The new biological and genetic screening procedures, though still undeveloped, would place responsibility on the individual at risk, and upon society to determine whether specific groups should be allowed to place themselves at risk.

### *ACGIH Standards and Industry*

It is easy to document the influence of industry, and of industry consultants in ACGIH. Because of this, and because labour has been a reluctant participant, it sometimes appears that ACGIH is an industry initiative and that its standards are biased. The fact that ACGIH standards are adopted by virtually all regulatory agencies does not deflect this criticism, but rather serves to raise questions about the biased nature of regulatory standards. Yet, ACGIH officials had a very different perception of the relationship between ACGIH and industry. ACGIH members vigorously denied that they were biased towards industry. In fact, many saw ACGIH as an industry "watchdog". It is important to examine the relationship between ACGIH and industry from their perspective.

Industrial hygienists are professionals in a specific sense of the word. They are often – though not necessarily – certified. They are often – but not always – trained in professional schools. Generally they act as middle management employees in large corporations (smaller companies do not employ hygienists, simply because of the cost). As employees, they have a mandate to protect workers' health and safety, but their daily responsibilities involve them in close interaction with other members of middle management. They must develop budgets for equipment purchases, and defend those budgets within the sphere of corporate planning. They inspect the worksite, but as representatives of management, not unions. They are, if speakers at the AIHA conference are to be taken seriously, not encouraged to speak with workers or to interact regularly with workers' organizations. They report to management.[19]

This situation creates conflicts for hygienists, for they are both professionals and employees. They must protect workers, but act for management in relation to those workers. This conflict is discussed quite openly at the AIHA conferences. In AIHA educational seminars, lawyers urge industrial hygienists to get their own professional liability insurance and to speak with workers regularly as a way of limiting their liability. Industrial hygienists are also urged to be cognizant of their professional mandate, rather than of the demands of their employers.[20] The advice of these lawyers at the AIHA conference was questioned by the seminar participants, as it reasonably can be. These industrial hygienists are concerned about the implications of having "split loyalties". Reassured that they can and must act first as professionals, the participants in the legal seminars continued to raise questions about how to assert their professional status on the job, for most of their daily activity is taken up in conducting measurements and ordering equipment.

ACGIH is not really a professional association, although its members are usually industrial hygienists. ACGIH seeks to control industrial practices, but it can do so effectively only with the active participation of the industry it is attempting to control. Members of ACGIH are often themselves regulators, but their work is only sometimes supported by the existence of regulatory standards. They know that not all contaminants are subject to regulatory standards and that many companies choose to function without any industrial hygiene program at all. In effect, good industrial hygiene is still, in many cases, left to the discretion of the individual company.

Concentrating on the experience of contaminants in the workplace allows ACGIH to respond to the practical problems of industrial hygienists, and the difficulties such hygienists face in persuading management to co-operate with worker safety programs. At the same time, the TLVs serve to deflect legal responsibility from the industrial hygienist as a professional. It is not surprising, on closer examination, that ACGIH has yet to face legal action and that industrial hygienists are only now considering whether to purchase professional liability insurance.

The relationship between ACGIH and industry can also be seen by examining the controversy in ACGIH over the adoptions of the Short Term Exposure Limits, the STELs. In the late 1960s, ACGIH decided to develop a standard for short-term exposures to chemical contaminants because obviously such a standard was needed in the workplace. Proponents in ACGIH of the STEL believed that its use would permit greater stringency in chemical control. But data – even experiential data – were not available for developing each STEL. If STELs were to be introduced, some other method of deriving them had to be used. A committee member described the introduction of STELs, as follows:

> When I was away and not serving on the TLV committee, some of (its members) thought that some of the numbers were a bit too relaxed and therefore they adopted the STEL concept. It was their way of reducing exposure...but they were done in a mathematical approach. Stokinger had come up with a rules-of-thumb guideline for excursion limits, so that if you have a TLV...you could have an excursion below as well as above, as long as it averaged out. But what's the maximum of the excursion? ...So Stokinger had rough rules-of-thumb and the maximum excursion he would accept was three-fold for TLVs in a certain range, one and a half in another range, two-fold in a certain range.
>
> ... But you see hygienists kept saying, "does it mean you can never exceed that?" "Okay, you can. How much higher can you go?". And that was a guide for him.... Well, the STEL people (advocates of the STELs) came along and they say, "Okay, you can have them, but they should not be more than four in an eight hour day. And they can't be more than fifteen minutes, and you must still live with the TLV-TWA concept.
>
> And they applied the mathematical rule and put them in holus bolus and I think that caused a lot of trouble for the TLV committee. Because people did not understand them initially. (People said) "Are these short-term TWA's" Are they maxima?"...And that is why they took them all out. What does science say about carbon monoxide? Can I be exposed to 400 ppm for 15 minutes, once an hour?" So they are going to use scientific literature...not mathematical formulas.

The introduction of STELs was vigorously defended within ACGIH, as is evident in this excerpt from the 1977 annual conference report:

> When this particular subcommittee deliberated as a group to determine what the value should be, we listed various categories of reasoning, one of which was the excursion factor (that which is referred to as the mathematical formula), but several other criteria were the Pennsylvania standards – or a reevaluation of the documentation of the standards which made statements to the effect that at the given TWA, there were short-term irritation effects, so that it should have been more of a ceiling with the TWA.[21]

Why would STELs be introduced, defended like this and maintained for many chemicals for years, given the difficulty of arriving at them and the controversies they generated? The answer can only be our guess, for it relates, we think, to the ambiguous role that ACGIH plays in relation to industry and workers' protection, and the inherent difficulties of gathering scientific information to support the introduction of STELs. We have some of the correspondence between committee members from the relevant time period. Since it is incomplete, we can use it only as examples of a debate we think occurred and as evidence to support our interpretation. Nonetheless, excerpts from the letters we do have are worth reproducing here:[22]

> *Example one, 1979:* "After considerable reflection, I believe that the STEL concept is fuzzy at best and is industry oriented rather than worker oriented. In fact it would appear to provide a lesser margin of worker safety than the "C" designation...Although the Excursion value concept has not been thoroughly proven, neither has it been disproven. For the most part, it has withstood the test of time."

> *Example two, 1976:* "In its original concept, no exposure time was assigned to the "C" limit, and according to the present membership of the TLV committee, no definite time limit is intended now. For the short-term acceptable exposure limit, the TLV committee has proposed the Short Term Exposure Limit (STEL) with arbitrary restrictions, based mainly on compromise and standardization."

> *Example three, (from an industry consultant) 1977:* "The anomalies and inconsistencies which have been introduced in the columns of tentative values for STELs have damaged the reputation of the TLV Committee. The reckless application of excursion factors to define the limit for short-term exposures,...would impose needless hardship in complying with the recommended values."

> *Example four, (different industry consultant) 1977:* "We were concerned about the use of Excursion Factors to establish STELs and the creation of STELs for every item on the TLV list...Publishing a list of STELs based on Excursion Factors will result in the rules (of thumb) becoming "fact" and to their use in laws and regulation. It is not desirable to have STELs for every material.... It is my understanding that the STEL subcommittee was asked to reconsider all those suggested STELs which had been based on Factors. The subcommittee was to determine if any STEL was necessary, rather than list a STEL with every TLV. May we reconsider this issue at the next committee meeting."

> *Example five, 1978:* "Since almost all of the substances listed in the Notice of Intended Changes for 1978 do not have a suggested STEL, we have attempted to conjure one up for each substance, based on their documentation writeup *or* the Minutes of the April 1978 TLV Committee meeting."

> *Example six, 1979:* "I personally am opposed to the extensive use of ceiling values. In my opinion Herb (Stokinger) used this device to accomplish a de facto reduction in the TLV. As a manoeuver to overcome opposition to any lowering of the limit, this has been quite successful, but I prefer a less Machiavellian approach. NIOSH and OSHA appear to use similar tactics."

> *Example seven, (from another industry consultant) 1979:* "The concept of a short-term exposure limit equal to a time-weighted average value for eight hours is

unsound. Another point is that the ACGIH definition of a ceiling 'The value not to be exceeded even instantaneously' is impractical in application to a real work situation."

*Example eight, again in 1979:* "It is intended that ceilings and STELs represent concentrations which should not be exceeded at any time. When there are short-term fluctuations in levels, inherent in the process, brief excursions above "C's" and STELs may be permitted as follows...The numbers (given in the letter) may be changed according the the wisdom of the Advisory group, if approved by the Committee."

*Example nine, shortly thereafter, in 1979:* "In looking at the definition of STELs, they are either much more restrictive than they appear, or else have a big loophole. With (chemical name), STEL 20 ppm, for example, one can have four fifteeen minute exposures per day at, but not exceeding, 20 ppm. How about exposures at 19, 15, or 11 ppm? Are these restricted in the same way, or is there no restriction as long as the TLV is not exceeded? In the former case, there could in practice be only four fifteen minutes a day in which the TLV was exceeded at all. In the latter, a worker could be at 15 ppm for nearly six hours, provided the other 2+ were exposure free; or even 19 ppm for a shorter period. The enclosed suggested modification of the ceiling and STEL definition is designed to overcome the objections of the AIHA.... These are only my suggestions. Any alternate way that effects the purpose is o.k. with me."

It is easy to see how TWAs might be used and still workers' health endangered in the workplace. A few peaks of chemical exposure can be lost in the averaging of such exposures over a long period, yet these peaks of exposure can cause considerable damage. The excursion limits, upon which STELs were first based, are themselves subject to controversy, but without these excursion limits and ceiling values, no limits to peak exposures would exist. STELs were simply an attempt to be systematic – in light of insufficient data, it appears – in tying down some of the variables and conditions that might otherwise turn TWAs and excursion limits into meaningless standards. However adequate, STELs represented a move towards greater stringency in the control of contaminants.

At the same time, the data upon which a STEL could be based were necessarily suspect, since toxicological studies are not designed to produce these specific measurements of harm, even for animal species. Extrapolation would have been required, even if sufficient data were to have existed. Establishing STELs by virtue of a rough "rule of thumb" was consistent with what ACGIH had done in its past, and consistent with the history of TLVs, which had been established more or less in the same fashion in the early days of the organization. It was logical to assume that STELs, like other TLVs, could and would gradually be backed up with proper data.

What ACGIH could not count upon in 1975 and thereafter, was that the early unanimity that greeted the first TLVs would still exist for STELs. The environment for standard setting was considerably different in the mid- and late 1970s than it was in the 1950s. A rule of thumb would suffice in the 1950s; it could easily be dislodged by industry and other criticism in the 1970s. The expectations of the scientific basis for standards had increased considerably in the interim. More important, the relationship of industry to the standards themselves was changed by the introduction of regulatory standards and litigation arising from them.

In the period between the introduction and documentation of the first TLVs and the introduction of the STELs, a number of other organizations had entered the business of standards. Toxicological research had advanced considerably, and toxicological data were seen – rightly or wrongly – as providing the only genuinely scientific basis for standards. Since regulators, not ACGIH, were primarily dependent upon toxicological assessments, regulatory standards were seen to be more scientific than those of ACGIH. ACGIH was not itself in court, but regulatory standards were being challenged regularly through the courts.

As the first organization systematically developing standards, ACGIH could offer a quasi-regulatory service to all. Once a coherent (albeit not necessarily adequate) body of regulatory standards existed, as they did after 1970, the environment for standard setting changed. ACGIH, other standard setting bodies and regulatory agencies were in competition. The working relationship between ACGIH and industry was thus changed. ACGIH remains dependent upon industry today and still conceives of itself as the guardian of workers' safety, but the equilibrium between these two pressures has been upset.

We found that it was much harder to discuss the relationship between ACGIH and industry today than the historical documents suggest was true in an earlier era. The concern about the relationship between ACGIH and industry that was expressed in its early phase still exists, but some ACGIH members are now vociferous in their defence of the organization as independent. ACGIH standards are scientific, it is argued, because experts are involved in setting them. The scientific aspect of ACGIH standards is seen to mitigate the possibility of an industry bias.

### *ACGIH Standards and Public Policy*

We can think of no better term to describe the relationship between various standard setting organizations – governmental, voluntary and consensus – than the faddish term "networking". Each of the standards bodies in various countries (including government departments and agencies) has its own mandate, structure and organization. Each follows its own procedures (some scanty, some well developed) for the assessment and adoption of standards. Each assigns to the resulting standards its own regulatory status. OSHA is very different from EPA, although both are American government agencies. Both are different again from ACGIH, ASME, ANSI, AIHA, all of which function within the United States.

These initials are confusing, but the nature of each organization and the differences among them with respect to the setting of standards (not their regulation) are really mainly of interest only to the insiders. A similar list of acronyms can be generated in other countries and at the international level. We have spelled out the titles for ourselves, and located some of the differences between these organizations. What is striking about them, however, is not their differences, but the manner in which standards decisions in one body influence another. Decisions about standards taken in one organization affect decisions taken in the others. Information about where a standard should be set that is generated in one organization is used by the others to revise their standards. A change in a standard by one organization will provoke a reassessment on the part of other organizations with respect to the same standard. Comparisons among the groups about specific standards are commonly made in the meetings of each organization.

Sometimes, this ill-co-ordinated movement of specific standards from one body to another is deliberate, as was the case when OSHA adopted the ACGIH standards after 1970, or when British Columbia adopted TLVs and converted them into regulations. Sometimes one sees more amorphous patterns of influence, or the cross-over of personnel from one standards body to another. Sometimes the jurisdictions of the various bodies overlap in a critical manner. EPA, OSHA, USDA, FDA and ACGIH all deal with pesticides in the United States (this list can be expanded by examining pesticide standards in other countries, of course, and acknowledging that most pesticides move in international trade) but under different mandates. At the very least, the standards set by any one organization act as reference points for the decisions taken by another group.[23]

Other factors appear to influence the adoption of ACGIH standards specifically, beyond those which we have already described. Not all countries have the full regulatory apparatus for the development of standards that is characteristic of the United States (however adequate the American system might be). In a country with limited resources for regulation or a minimally developed regulatory process, the adoption of ACGIH standards represents a first, and significant attempt to generate control over contaminants in the workplace. In this context, ACGIH standards are much better than no standards at all.

Many countries also maintain a workers' compensation scheme, a state-run insurance system for injuries to health and safety on some jobs. Such compensation schemes cannot operate without standards to determine their rates and the appropriateness of their regulations and their settlements. Compensation boards are only rarely equipped to develop a full range of standards through a local process of scientific assessment. With few resources for developing their own standards, the compensation boards are dependent upon standards set in other jurisdictions.

ACGIH is in a unique position to provide the standards that these countries, their agencies and insurance companies require. It is both a voluntary organization and a public one, in the sense that its deliberations are public and its members mainly government officials. At the same time, ACGIH is not a governmental organization, so national jurisdiction is not compromised by the adoption of ACGIH standards. ACGIH standards are also accompanied by documentation. Users of ACGIH standards can confidently assure their constituencies and those to whom they report that local assessment (of this documentation) has been conducted by simply reviewing the ACGIH documentation in light of their own local experiences with the chemical. The fact that ACGIH standards are based on experiential, not just toxicological data makes their adoption with reference to local conditions seem quite legitimate.

ACGIH standards are so widely adopted simply because they exist, because they are seen to be politically neutral, experientially based, reasonably scientific and fairly comprehensive. The adoption of ACGIH standards can proceed without any formal or even informal co-ordination or agreement between national governments. The TLV booklet is available to any purchaser. As guidelines, not regulations, ACGIH standards can be taken out of context, and adopted without reference to the significant differences in regulatory philosophies and processes that exist in various jurisdictions.

This is not intended to be a defence of ACGIH, nor has any of the foregoing analysis of their activities intended to provide a basis for a moral evaluation of ACGIH or its standards. ACGIH is often very frank about its limitations, too frank to require any defence from us. Our description is intended to illustrate how standard setting works in practice; our example of ACGIH as a standard setting organization is chosen because we think it is a very important one.

Before we evaluate ACGIH, or indeed discuss it further in terms of an analysis of the characteristics of mandated science, we need to see ACGIH in the context of other bodies setting standards and of the kinds of controversies that chemical standards generate. It is too easy, and fundamentally a flawed approach, to criticise ACGIH on the basis that their standards are insufficiently scientific, their process less than democratic and their activities unduly influenced by the companies that produce contaminants as part of their process of manufacture. ACGIH can legitimately be criticised on all of these grounds, but doing so tells us little about the real options for setting standards or for the development of an adequate mandated science.

Chapter Four

# Alphabet Soup
# Case Study Two: The Codex Committee on Pesticide Residues

In the post Chernobyl period, awareness of the existence of international standards organizations has increased dramatically. Like the international Red Cross, such organizations are seen to be a source of unbiased and scientific information about public safety. People are reassured when they are told that the radiation in the drinking water conforms to, or exceeds only marginally, the international standard for radiation. Everyone breathes easier if the levels of airborne radiation meet the international standard.

Should people be so easily reassured? That will depend upon how the international standards are set, the quality of the scientific information, and the definition of acceptable and safe that the international organizations use. About these issues, there is considerably less assurance. Indeed, it is almost as a matter of faith that international standards are accepted as sufficient for public safety.

We examined international pesticide standards and the Codex Committee on Pesticide Residues (CCPR). CCPR sets standards for the amount of pesticide residues on food. It is a technical committee of the Codex Alimentarius, a United Nations Commission dealing with food standards of all kinds. Codex Alimentarius has a number of standard setting committees similar to CCPR.

In fact, three United Nations organizations are involved in standard setting for pesticide residues, Codex Alimentarius, CCPR and a committee of experts, the Joint Management Committee on Pesticide Residues (JMPR). Codex is sponsored by the World Health Organization (WHO) and The Food and Agriculture Organization (FAO). Codex Alimentarius, not CCPR, is responsible for the actual adoption of standards and their publication, although the primary standard setting activity for pesticide residues is done by CCPR. The JMPR is responsible for the scientific deliberations used in setting pesticide residue standards, although it does recommend standards. The JMPR is independent of CCPR and reports directly to Codex. As an expert committee, it is considered to be very different from either Codex or CCPR. Codex, CCPR and JMPR work together, and the Codex system of standard setting can only be understood from the complex relationship among these three groups.

The Codex system is not the only process for setting international pesticide residue standards, nor is the JMPR the only international body that conducts scientific assessment for the purpose. The European Common Market has an elaborate – and somewhat similar – system of standard setting for its members. OECD plays a minor but important role in standard setting for chemicals including pesticides. Two other groups sponsored by the United Nations, the International Program for Chemical Safety (IPCS) and the International Association for Research on Cancer deal with pesticides, and indirectly with pesticide standards.

Among these organizations, Codex is the most important. More than one hundred countries are members of Codex. Only Codex standards are used throughout the world, and particularly in third world countries. The JMPR is relied upon by IPCS for most scientific assessments related to pesticide residue standards, because often only the JMPR has access to the information to be evaluated.

Codex standards are similar to ACGIH standards in that they are voluntary guidelines, not regulations. They can be adopted as regulations by different countries, with or without further assessment. Only in recent years has CCPR paid significant attention to whether its standards are adopted as regulations. As a consequence, Codex provides an illustration of how important voluntary standards can be.

Codex is important for our purposes for another reason. Pesticide residue standards are based on scientific and nonscientific information. In the case of Codex, the nonscientific information is primarily about trade. Codex formally recognizes the importance of standards for trade relationships, and it sets its standards accordingly. Its goals include the promotion of trade. Codex standards and the differing patterns of their acceptance throughout the world have a great deal of influence in trade relations. The relationships of interest in mandated science, the relationships between science and public policy for example, are clearly spelled out in the mandate for each of the Codex organizations.

Initially, we were impressed with Codex as a standard setting organization dedicated to furthering co-operation among nations, and to the health and safety of all citizens. To the extent that Codex develops safety standards that are used by many countries, and that it removes artificial impediments to trade, then the goals of the United Nations are also achieved. The research necessitated a reassessment. The decisions made by Codex and its committees are anything but symbolic in their importance. They have a direct effect upon the profit levels of major corporations. The trade relations created by Codex standards are likely to benefit some countries, and some interest groups more than others. Whether the Codex standards actually contribute to the objectives of the United Nations is a matter for investigation and debate. We have some doubts.

### The Codex Alimentarius

Although CCPR is of most importance for the study of mandated science, we will begin with the parent organization, Codex Alimentarius. International standards relating to food are set by a Commission within the United Nations called the Codex Alimentarius (Codex). The name decribes the organization and also its products, the standards themselves. Codex sets many different kinds of standards: standards for quick frozen foods, for infant formulas, for various food commodities, for food additives as well as for pesticide residues.

Codex was formed in 1958 as a European organization. In 1962, after some administrative and financial difficulties, the organization became the joint responsibility of WHO and FAO. Today, Codex is made up of members and associate members of FAO and WHO who have notified the Director Generals of either organization of their wish to be considered members.[1] Codex has a number of subsidiary bodies dealing with policies and with specific commodities, and CCPR is one of its six commodity committees. Within Codex, responsibility for different aspects of standard setting is taken by various host countries. In the case of CCPR, The Netherlands plays the host role.

The Codex mandate is linked to food standardization, to "make a major contribution to ensuring that safe, adequate and acceptable supplies of food are generally available throughout the world".[2] A Canadian official describes the Codex view of standardization as one ensuring: "the most efficient and least wasteful use of this precious commodity."[3] The general principles of Codex state:

> Codex Alimentarius is a collection of internationally adopted food standards presented in a uniform manner. These food standards aim at protecting the consumer's health and ensuring fair practices in the food trade. Their publication is intended to guide and promote the elaboration and establishment of definitions and requirements for food, to assist in their harmonization and in doing so to facilitate international trade.[4]

Several aspects of this mandate should be noted. First, Codex deals with possible dangers from the use of pesticides only from the perspective of the consumer. Pesticides used in industrial settings or on non-food products, and pesticides that do not leave residues on food that is consumed, fall outside the scope of interest of Codex. For example, the protection of the pesticide applicator or of those living near farms being sprayed are outside the mandate of Codex. Second, the link between food standards and trade is an explicit one. One of the main aims of Codex is the protection of "fair trade in food".

Third, Codex standards are intended to act as a "guide" and to "promote ... requirements for food". Codex standards are, in other words, only guidelines and not regulations, even if they are accepted and published with the formal agreement of Codex member nations. Finally, the Codex process is intended to be supportive of national regulatory systems, where they exist, and of the development of regulation ("the elaboration and establishment of definitions and requirements for food") where they do not. Codex is not a regulator, nor an alternative to regulation. It is, instead, a promoter of national regulatory activities.

The inclusion of trade and economic considerations is recognized formally in Codex. In its procedural manual, Codex states:

> It will be open to any Member of the Commission (Codex) to draw to the attention of the Commission any matter concerning the possible implications of a draft standard for its economic interests .... In considering statements concerning economic implications the Commission should have due regard to the purposes of the Codex Alimentarius concerning the protection of the health of consumers and the ensuring of fair practices in the food trade, as set forth in the General Principles of the Codex Alimentarius, as well as the economic interests of the Member concerned.[5]

Codex reports through FAO and WHO, but FAO carries the administrative and budgetary responsibility for its activities. Associate Members or Members of FAO and WHO can be observers, as can nations which are not members of the United Nations. Member delegates are drawn from a variety of different governmental departments or agencies in their own countries, including ministries of public health, governmental offices for "Specifications and Measurements", ministries of agriculture and land, industrial testing and research centres, ministries of rural development, and of international affairs. These departmental affiliations provide an indication of the very different ways in which Codex activities are viewed by its participants.

All Codex meetings are open to all Members, however, and international organizations – including industry or trade associations – may be granted observer status if they have such a status with FAO or WHO. Codex meets yearly, and the quorum for its meetings can be as low as 20% of the total membership, or 25 Members. If a regional issue is involved, the quorum is one third of the Members from the region. It would be a mistake to think of the general meetings of Codex as regionally balanced. Member countries and observers are responsible for their own expenses, except if special allocations are made by Codex for some of the preparatory work. Countries with the resources and desire to participate actively are obviously in a privileged position to do so, given the decentralization of the incidental costs connected with membership. This has created a gap between participation from the developed and underdeveloped countries.

Of the more than 120 Member countries, 58 were in attendance at the 1983 meetings.[6] The level of national participation varies greatly, as does the proportion of nongovernmental (usually industry) members of each national delegation.

Even taking the fact that countries must pay their own expenses and that these expenses vary considerably (the meeting was in Rome), some interesting patterns emerge from this list, which is typical of other Codex sessions. Eastern bloc countries do not appear to be very active in Codex, with the possible exception of Hungary. This observation was confirmed by people we interviewed. As might be expected, Third World countries are represented by fewer delegates, and industrial countries have larger delegations. There appears to be no relationship between either dependence upon agricultural trade (or level of agricultural production) and participation in the meetings. Countries determine the size of their own delegations. Most significantly, different countries follow very different policies – for historical reasons, it is claimed – with respect to the inclusion of delegates from industry.

TABLE 5

*Attendance at the 1983 Codex Alimentarius*
*Proportion of governmental delegates noted in brackets*

| Countries with 1–3 delegates | Countries with more than 3 delegates |
|---|---|
| *Eastern bloc:*<br>USSR (1/1), Yugoslavia (1/1),<br>Poland (1/1), Czechoslovakia (1/1) | *Eastern bloc:*<br>Hungary (5/5) |
| | *North America*<br>U.S. (26/7), Canada (5/4) |
| *European:*<br>Portugal (3/3), Ireland (3/3) | *European:*<br>U.K. (7/6), Switzerland (8/2),<br>Sweden (5/5),<br>Spain (16/16), Norway (6/6),<br>Netherlands (10/8), Italy (37/34),<br>Greece (5/4), Germany (8/6), France (18/5),<br>Finland (6/6), Denmark (12/6),<br>Belgium (4/4),<br>Austria (4/3) |
| *Latin America:*<br>Nicaragua (1/1), Equador (1/1),<br>Bolivia (1/1), Argentina (1/1) | *Latin America:*<br>Brazil (8/4), Cuba (10/10),<br>Mexico (5/3) |
| *African and Middle East:*<br>Turkey (1/1), Tanzania (1/1), Senegal (1/1),<br>Saudi Arabia (1/1), Nigeria (1/1), Liberia (1/1),<br>Kenya (2/2), Ivory Coast (2/2), Iraq (2/2),<br>Iran (2/2), Ghana (3/3), Egypt (1/1),<br>Central African Republic (2/2),<br>Cameroons (2/2), Algeria (2/2),<br>Afghanistan (1/1), Cameroon (2/2) | *Africa and Middle East:*<br>Tunesia (4/3), Gabon (4/4) |
| *Asian:*<br>Bahrain (1/1), Philippines (2/2),<br>Pakistan (1/1), Korea (3/3),<br>Dem. Rep. Korea (3/3) | *Asian:*<br>India (6/6), Thailand (8/8),<br>Japan (5/5) |
| *Other:*<br>New Zealand (3/3), Australia (3/3) | |

The observer organizations included twenty-five organizations that could be readily identified as industry or trade associations, and twelve others with some standard setting activities. Nongovernmental participants comprised a significant proportion of the people present at the meetings, although formally, of course, only Member countries have a vote.[7]

The annual meetings of Codex provide the opportunity for a brief discussion of the standards recommended by the technical committees and vetted by national governments. If the minutes are an accurate guide, the discussion is limited to the presentation of national and official positions with respect to the standard under debate. In a few cases, scientific matters are raised, but generally the debate focusses more directly on issues of national interest and trade. Other issues are addressed: the membership of Codex, the role of developing or third world countries within Codex, the acceptance of its standards, the activities of other international standards or related organizations, regional reports, and the development of a code of ethics for international trade in food (relevant in the debate about baby formulas, for example). On each issue, and with respect to each standard under consideration, individual delegates, and observer organizations briefly express their official positions.

Codex sets standards through an eight stage procedure (the first two stages are usually omitted) that allows for several rounds of consultation, primarily through CCPR, with national governments before a standard is adopted and published. As one official who has been a delegate to several meetings suggested:

> If most of the countries are objecting to the proposal for one reason or another, it would be sent back ... with reasons for objections. If most of the countries are accepting it, it will probably be proceeded with. If a few countries are objecting for a specific reason, the reason might be addressed to see if it can be resolved.

Codex standards take many years to emerge from the process. For example, although the organization had been in existence since 1965, no international Codex pesticide residue limits were ready for publication until 1972.[8] It normally takes about eight years for a standard to go through the full eight stages and to be published by Codex. We were told that at any time, one hundred chemicals might be in the system. Amendments to standards follow a similar staged process of approval. Standards or amendments for pesticide residues can be adopted more quickly, however. If the matter is of some urgency and governments agree, three of the steps can be omitted, and consequently, a second round of Member consultations will not take place.

A distinction is made between adoption of a Codex standard and its acceptance. Adoption means that a standard has an official status under Codex. Acceptance means that countries have decided to use the Codex standard as part of their own national regulatory systems. At the eighth stage of the Codex process, the individual governments have the opportunity to *adopt* or reject the proposed standard. Countries can – and do – vote for the *adoption* of a Codex standard and then later fail to *accept* it as their own. Not all Codex standards have been accepted by all member countries by any means.

Codex provides a number of different categories for national acceptance of its standards. Thus member countries have considerable discretion about their level of commitment to the Codex standards. Full acceptance of Codex standards means "that the country concerned will ensure, within its territorial jurisdiction, that a food, whether home produced or imported...will comply." Limited Acceptance means that the country involved "undertakes not to hinder the importation of food which complies" and that the country does not impose a more stringent standard under Codex than is applied domestically.

As well, there are several categories of full and limited acceptance. Target acceptance means that the country has indicated its intention to give full or limited acceptance after a stated number of years. Finally, there are several categories of nonacceptance, indicating varying degrees of co-operation (or non-co-operation) by Member countries with Codex standards.[9]

Codex is a large organization, and pesticide residue standards are a minor portion of Codex activities. It is not surprising that the detailed evaluations of the scientific and other information upon which the proposed standards are based are not discussed at any length in Codex meetings. Instead, these are a matter of concern for the national delegates, who submit complaints and recommendations in writing to the Secretariat, and for the technical committees that are mandated to provide greater scrutiny of the standards than Codex can provide.

## The Codex Committee on Pesticide Residues (CCPR)

In the area of pesticides, this technical committee is CCPR. CCPR is a group of particular interest for a study of mandated science. Like Codex, it draws its membership from national representatives. The trade and national interests of standards are dealt with explicitly by CCPR in the selection of pesticides for standard setting and in the nature of its delegates' mandate. But unlike Codex, CCPR members are often working scientists or technically trained individuals. They view themselves – or at least explain their activities – as relating to health and safety. It is CCPR that brings together the scientific assessments and their implications for standards. It is CCPR, not Codex, that adopts the numbers - the standards – initially, and CCPR reviews any comments from Codex Member countries with respect to the standards before changes are recommended.

CCPR held its first meeting in 1966, and it meets annually. Its terms of reference have been widened over time. A CCPR working paper provides an historical view:

> Its first approach was largely commodity oriented and the first session was devoted to one group of commodities (cereals and cereal products) and the relevant pesticides. It soon became apparent that a pesticide oriented approach, covering all relevant commodities, would be more realistic, and the subsequent programme was, therefore, based on lists of pesticides selected according to agreed priorities.[10]

Some work has been done to develop operating principles for sampling, and codes of practice. CCPR has now addressed the problem of unintentional contamination, pesticides entering the food chain from sources other than their application.

Forty-three countries attended the CCPR meetings in 1978; forty-six in 1983 and 1984; forty-one in 1985. Membership in CCPR has stabilized. Most countries send smaller delegations to CCPR than to Codex, yet the proportion of nongovernmental members on the official delegations in CCPR appears to be higher. The following countries send more than three delegates:

TABLE 6

*Delegates to CCPR (1985)*

*Proportion of Nongovernmental Delegates to Total noted in parentheses*

| | |
|---|---|
| Australia (4/1) | Germany (11/1) |
| Belgium (4/2) | Netherlands (11/4) |
| Cameroon (4/0) | Switzerland (6/3) |
| China (4/0: three from export corporations) | Thailand (4/0) |
| Finland (5/0) | U.K. (9/1) |
| France (5/4) | U.S. (14/4) |

In 1985, among the 131 individuals representing national groups, thirty-one individuals were consultants from industry.[11]

As a technical committee of Codex, CCPR is also mandated by and responsible to FAO and WHO. CCPR exists within the sometimes uneasy relationship between these two branches of the United Nations with their different mandates. With respect to CCPR, we were told that WHO is most active in relation to the expert committee (the secretary is from WHO) that advises CCPR. FAO provides the background information with respect to the agricultural practices influencing the safety of pesticide use. Neither WHO nor FAO maintains a large staff to deal with CCPR or pesticide standards, however.

We tried to estimate what proportion of delegates came from FAO and WHO related departments. We assume that representatives from agricultural ministries would focus directly on the benefits of pesticides, or at least their economic role in food production. As such, their orientation would be to the trade issues, not to the health and safety matters that are the primary mandate of CCPR. The proportion reflects an imbalance in CCPR membership towards trade issues, as Table 7 illustrates.

No farmer, labour, environmental or consumer groups were represented among the delegates or observers to CCPR meetings in any of the years we studied, but the trade organizations were very well represented. In 1985, there were forty-two observer organization representatives, including six from other standard setting or international organizations and thirty-four individuals from GIFAP, the international organization representing pesticide manufacturers. GIFAP members were represented on a few of the national delegations as well.

TABLE 7

*Functional Affiliation of Delegates to 1985 CCPR Meetings (Governmental, consultants and company delegates included)*

| | |
|---|---|
| Agriculture | 53% |
| Health or nutrition | 27% |
| Other governmental departments | 9% |
| Other | 11% |

Of the thirty-four GIFAP representatives, fourteen were from the USA. By virtue of their position in the alphabet, the official delegation from the United States was seated at the back of the room, immediately in front of the GIFAP observers. Although American and GIFAP interests no doubt diverge on a number of points – countries such as UK, France and Japan are represented in the GIFAP group – it seemed like the last three of the eleven rows of participants constituted a bloc of like-minded individuals. This perception, real or not, was shared by some of the delegates with whom we spoke, and certainly is problematic for CCPR.

CCPR meetings are much like those of Codex, in spite of the differences in their mandate and membership. Discussion from the floor in the general meeting is mainly limited to delegates presenting their official positions, although some debate does occur. A large proportion of the discussion is given over to a case by case consideration of specific standards for individual food items. Most countries do not speak on most issues.

In both Codex and CCPR, the issues are officially decided by majority vote, but formal votes were not taken in the CCPR meeting we attended. Instead, the Chairman used a modified consensus procedure, in which both delegates and observers signalled their concerns or agreement from the floor and the Chairman "read" the intent of the meeting and noted the decision. This consensus procedure is considered feasible in the case of CCPR because of the type of decisions it makes. Recall that a CCPR recommendation goes through several stages of consultation before it becomes a Codex standard. Dissatisfied parties have more than one opportunity to influence the outcome. And even when a decision is taken by Codex, member countries are not committed to accept the standard unless they choose to do so. Decisions taken by CCPR are only decisions in a limited sense of the term.

There is one area where CCPR does make decisions. The CCPR sets the priorities for chemical assessments. It relies upon its priorities sub-committee for recommendations. The official criteria for placing a chemical on the CCPR priority list (and thus, as we shall see, on the JMPR list) are that the compound must:

(a) be available as a commercial product
(b) not already have been accepted for consideration
(c) result in residues in or on the food commodity
(d) affect international trade to a significant degree
(e) be a matter of public health concern, creating or have the potential for creating commercial problems.[12]

Requests for consideration of a chemical to be placed on the priority list are sponsored by countries, and the manufacturer is listed. They are accompanied by a description of the uses of the pesticide and commodities moving in trade, a list of the countries where the pesticide is registered, and the existing national standards. The discussion in the sub-committee revolves mainly around the availability of data, which seems to be the main determinant of whether a chemical will be given priority for evaluation. As a consequence, it is not surprising that industry observers play an active role in the priorities sub-committee meeting.

## The Joint Management Committee on Pesticide Residues

To understand how CCPR – and indeed Codex – operates with respect to its specific standards, it is necessary to add a third organization to the list of participants, the Joint Management Committee on Pesticide Residues (JMPR). The relationship between Codex and CCPR is easy to understand, for CCPR is simply a technical committee of Codex. The relationship between JMPR and either Codex or CCPR is more complicated. JMPR is an expert committee, created before CCPR, and responsible to WHO and FAO. Its terms of reference are more general than those of CCPR.[13] JMPR is the committee which reviews the scientific evidence with respect to pesticide residues.

JMPR publishes documentation and a series of recommended standards for the consideration of Codex – in practice, for the consideration of CCPR. JMPR is officially independent of CCPR, although CCPR sets the priorities for its assessments. JMPR can be asked to reconsider its assessment and recommendation at the request of Codex or CCPR, but JMPR is intended to be independent of considerations of trade, national interest or the specific mandates of its parent organizations, FAO and WHO. JMPR decisions constitute the advice of a WHO expert committee, but they are not recommendations of WHO. Little wonder the relationship is complicated.

The JMPR itself is made up of two panels of experts, each chosen by the two parent organizations, WHO and FAO. Members are appointed on an ad hoc basis. These experts are appointed from "panels" or lists of experts maintained by the parent organizations, although the secretary of JMPR plays the major role in selecting the WHO experts. At the 1984 session of CCPR, it was noted "that every JMPR was different from the others and might come to different conclusions," and that "after the last day of the meeting, these people are not experts any more." Some carryover of appointments is encouraged, and indeed many of the experts participate for several years. It is unlikely that there is much difference in the JMPR from year to year. The JMPR issues a single decision that combines the evaluations of both of its own panels of experts.

The resources for the JMPR are meagre. The number of people appointed has remained constant throughout the years of JMPR's operations, although the workload of the JMPR has increased significantly. The JMPR does not have its own staff, other than a very small secretariat. It has no resources for scientific assessment, other than those contributed by the experts on its panel or its temporary advisors and the expertise of its Chairman. Its budget is used for travel. The JMPR is highly dependent upon the resources of its experts, and the expertise they bring to the committee.

Some countries are more active participants in the JMPR process than others. This activism reflects a number of factors, two of which are the sophistication of the regulatory capability in the Member country, and the willingness of countries to contribute the time of their officials. The level of activism also reflects the cost of participation, which can be considerable, given that individual members are responsible for preparing and evaluating documentation before they attend the meetings. JMPR is highly dependent upon the goodwill of various governments (although some members are academics, not regulators).

In spite of JMPR's reliance upon the good will of governments, we were told that national interests were "completely irrelevant". JMPR members are supposed to act "as international citizens", and "(they) are not coming to serve their own national interests". Seconded from government departments or themselves regulators, JMPR members are expected to function as if their regulatory experience and the national priorities of their government had no bearing on their recommendations. In other words, they are expected to insulate themselves, by virtue of their status as scientists, from the national or regulatory implications of the recommendations they make, even if they are also regulators.

The developing countries are represented as members of the JMPR expert panels, as they must be by the rules governing United Nations organizations, but we were told that their participation has been limited. Someone, usually a national government, must pay for their expenses in precious hard currency. This situation was commented upon by several people we interviewed who are active in international standard setting. One said:

> Here we get into a very difficult situation, because we are under a rule, very strongly enforced, that the committee has to have geographical distribution. And having only provision for six members, we have to pick up one expert from each region of WHO...So that's one strong limitation.

The JMPR normally meets annually, but a chemical might not be examined until two years after it has been added to the list of JMPR priorities, and conceivably later, if the JMPR fails to complete all the evaluations for chemicals already on its list. Typically, the JMPR might review thirty-five chemicals in its ten day working session each year. Six new chemicals might be added to its list in any session, and a number of other chemicals will be re-evaluated. JMPR meetings are not public, nor are industry observers permitted, although representatives from industry meet with FAO officials, and with the WHO secretariat.

Temporary advisors extend the resources of the JMPR. The temporary advisors are individuals chosen by the Secretary of the JMPR or by FAO to develop the documentation on particular pesticides. They serve at the pleasure of the Secretary or of FAO. They are not paid, and their participation is possible only if they can be seconded from their regular job, usually with a national regulatory organization. The temporary advisors sign an "ethics statement" that the information they receive will not be used for national purposes.

The process of assessment and standard setting in the JMPR is as follows. Draft documentation concerning a pesticide is prepared by either a member of the JMPR, a temporary advisor or the JMPR Secretary. The temporary advisors then receive and review the data, including the toxicology as summarized by the Secretary of the JMPR. This part of the process is very much like that of ACGIH, except that – to the best of our knowledge – industry consultants do not themselves prepare the original documentation in JMPR, as they do in ACGIH. According to an industry spokesman, however, the JMPR experts and its temporary advisors consult extensively with industry in preparing the draft documentation. The members of the JMPR told us that their contact with industry is limited. This is a sensitive issue, as one might imagine. The situation was described to us as follows:

> The implication is that the reviewers who do the work for WHO lack credibility and integrity, when it is said that we can't allow you to talk to industry because they will compromise you. I would find that insulting if I were a reviewer. These guys are pros...they don't need someone else to say you be careful and don't talk to industry. And because we have the data..its not a matter of (us) being unwilling to give anyone the data.

The defensive tone of this interview suggests to us that consultation with industry does take place when the draft documentation is being prepared.

Once the draft documents are prepared, they are reviewed first by a member of the JMPR and then brought to the JMPR meeting for debate and decision. At the end of its meeting, the JMPR publishes a series of recommendations for standards for those compounds for which it has adequate data. It also publishes extended evaluations of the chemicals it has reviewed. In recent years at least, the more detailed evaluations have been published after the CCPR has taken action on the JMPR summary recommendations.

The sources of JMPR data were described in its working paper in 1978, as follows:

> Although the individual members of the JMPR are expected to undertake their own researches of relevant literature in conducting their evaluations, Joint meetings have to rely to a great extent on information assembled and made available by industry and governments. Requests for such information are distributed widely *through the procedures of CCPR* (emphasis added), the Industry Association (GIFAP) and other means. This is done by the distribution of standard circulars and the secretariat of the JMPR has little or no capability to communicate and solicit information from specific sources.[14]

The JMPR is unlikely to obtain data on a specific pesticide unless countries and companies support its assessment or re-evaluation, since compliance with the JMPR requests for data is entirely voluntary. Deregistration of a pesticide in one country (for example, the banning of DDT) would not necessarily put that chemical on the agenda of the JMPR (or alter the CCPR decision with respect to it). It would be appropriate, then, to view the JMPR activities as largely data-driven, and the availability of data as a means of controlling JMPR's activities.

When compared to the pattern of re-evaluations of ACGIH, JMPR evaluations appear to be more frequent. Many chemicals have been examined more than once and many, often. The chemicals re-evaluated most often are captan (8 times), carbaryl (ll) cyhexatin (8), DDT (8), heptachlor (9), lindane (ll) and malathion (9). These are all chemicals that are easily recognized for their controversial status.[15] The JMPR can, but does not often take the initiative on a pesticide that is not already under assessment, unless it is in response to a deadline previously set in a JMPR evaluation or without direction from a Member country and/or CCPR.

## The Three Organizations

The three United Nations organizations creating standards for pesticide residues on food – the initials in the alphabet soup – are: Codex, its technical committee CCPR, and the expert committee, JMPR. These three bodies operate with different mandates, but in some senses, they are engaged in the same task. As we have indicated, there is extensive reliance of each organization upon the others, but each has its own agenda and value orientation.

The scientific material serving as the basis for a decision on a pesticide residue standard is filtered through all three committees, supplemented informally en route by the information of experts, industry and/or Member countries. By the time standards are set in Codex, the scientific information that has entered JMPR (whatever its merits as science) has passed through a double process of further assessments by bodies with profoundly different mandates from the JMPR. In the approach of the JMPR scientific values prevail. Nonetheless, the JMPR decisions are re-evaluated many times before they are reflected in the more politically oriented decisions of Codex.

National interests play a critical but different role in each round of the assessment. In the JMPR, national interests are all but invisible, operating at the level of structural constraints rather than influencing specific decisions. At CCPR meetings, national interests can be discerned, but the discussion about them is confined to the trade implications of different decisions, broadly considered. National interests influence the comments made at CCPR sessions by technical personnel with respect to health and safety considerations, but this is never openly acknowledged in the CCPR debate itself, or anywhere except in informal – and usually off the record – conversations. National interest considerations are openly acknowledged in Codex, although they are supposed to be tempered with concern for human health and safety and with the desire to promote trade generally.

It is useful to construct a schema of the relationships between different kinds of assessment in the the Codex system:

TABLE 8

| | | JMPR | CCPR | CODEX |
|---|---|---|---|---|
| | high | scientific issues | health and safety issues | trade and national interest |
| priorities for assessment | | Health and safety issues | trade issues | health and safety issues |
| | | | scientific issues | |
| | low | (trade and national interests) | | (scientific issues) |

In the case of JMPR, the discussion is about scientific information and health and safety issues. Trade and national interest considerations are submerged, neither recognized formally nor acknowledged in discussions. Participants in JMPR are aware of the consequences of their assessments for trade or economic issues, of course. Many of them are regulators who deal with economic assessments regularly in other contexts. In Codex, trade issues are given primary importance and national issues are presumed to play an active and legitimate role. CCPR sits at the juncture of these two organizations, combining some aspects of both.

In general, we found the relationship among the three organizations to be an uneasy one. Members of CCPR and JMPR each stressed the differences between their activities and those of Codex. There is much evidence of friction. For example, one of the WHO appointed officials said:

> No, please, no. It (the JMPR) is not part of Codex. The liaison between JMPR and Codex is that JMPR is a scientific body and Codex is a political body...It has been a fight to keep the independence of JMPR.... Sometimes they want JMPR to re-examine what the JMPR has done. Reinterpreting the data, but we will say no...How can you be part of Codex if you are acting in an advisory capacity...otherwise the advice will come from within.

The fact that CCPR sets priorities for JMPR creates considerable difficulty between the CCPR – whose members are scientists or technical people in their own right – and JMPR. The JMPR assessments often fail to satisfy the CCPR. Lacking other information, CCPR members stress the fact that they are often required to accept limited documentation for the purposes of making CCPR decisions. The dissatisfaction was described to us often, and it underscores a statement in the 1978 CCPR report:

> There have been long delays in the issuance of JMPR documents. There has been great difficulty in maintaining JMPR meetings on an annual basis, and it has not proved possible for the JMPR secretariat to seek out or to assemble data systematically or to arrange for its ready retrieval.[16]

The following statement from a JMPR working document in 1978 describes the general picture today as we saw it:

> The recommendation coming from the meetings (of JMPR) can only reflect the situation as available in the documents presented and the present knowledge of the attendees themselves. As proposals for MRL's (standards) have to be prepared at sessions, it is not possible to wait for details of other information that may be known to exist. Nor is it always possible to provide a very firm judgement as to whether a particular use represents "good agricultural practice" or whether a certain residue level affects international trade so as to create problems. Indeed, the discussion of such matters at the CCPR frequently reveals that information on such matters were not made available to the JMPR. The number of such cases should decrease if the capability of the secretariat to solicit or otherwise collect information were supplemented. But it is not envisioned that the JMPR can pass final judgements on such matters. Consequently, there will continue to be a need for adjustments to be made by the CCPR or for requests for re-evaluation to be sent back to the JMPR based on the availability of new information.[17]

A number of suggestions to change the JMPR have been made to take account of these problems, including one to merge the CCPR and the FAO functions now conducted through JMPR. As of 1985, there was no evidence that any significant change had taken place. CCPR reviews the JMPR recommendations. CCPR members have their own views on "good agricultural practice", and rely upon their own judgements in developing CCPR recommendations. "Adjustment of JMPR recommendations" (the term comes from the 1978 report) by CCPR after they have been received from the JMPR can be done.

## The Standards

With an understanding of the general structure of the Codex and JMPR, it is now possible to examine the standards themselves. The primary standard of Codex is the maximum residue limit (MRL), which is set for each product and pesticide.[18] Given the number of food products for which a single pesticide can be used, the number of MRLs is very great. This large number of MRLs should not hide the fact that there are no standards for many pesticide residues.

A maximum residue limit is the amount of residue of a pesticide on some portion of a food crop, considered to be tolerable toxicologically and consistent with good agricultural practice. Since this definition raises more questions than it answers – what is tolerable, which portion of the food should be considered and what is good agricultural practice? – it will be useful to show how a MRL is created. Unfortunately for our purposes, this is not an easy task.

In our research, we found a number of inconsistencies in the descriptions of Codex standards and the method by which they are derived. Superficially, the process appears uncomplicated, but these inconsistencies cloud the picture. After much frustration, we were actually relieved to discover that JMPR and Codex use somewhat different definitions for even such basic terms as "pesticide residues",[19] for it indicated to us that the problem is endemic to Codex, not our research. To resolve the confusion, we will provide a simplified discussion of the standard setting procedure used by Codex, and discuss the inconsistencies afterward.

The process begins with an ADI, a standard for an "acceptable daily intake" of any pesticide residue. The acceptable daily intake is related to the pesticide – not the food on which the pesticide residue can be found. To develop an ADI, a threshold or "no observed effect" level is determined for the pesticide from a review of the toxicological literature. A "no effect level" is the amount of pesticide residue that will produce no apparently serious and harmful effects on test animals in toxicological testing.

Assuming that a no effect level exists for the pesticide residue in question, a safety factor is then used. The amount of residue considered "acceptable" is reduced proportionately to the danger imposed by the chemical in question. In other words, the number indicating the lowest level of a pesticide that will cause toxic effects in a laboratory animal is multiplied by another number, representing the degree to which JMPR members feel that caution should be used in applying the toxicological data to humans. The safety factor is designed to take account of human exposures and harm. It is a number reflecting the judgement of the experts involved in making the decision. This "no effect/safety factor" method is used by several countries in registering pesticides.

The ADI does not actually reflect what people eat, their "daily intake". [20] It is regularly exceeded in some countries, and is generous for many. It is not a measure of the pesticide residues consumed by people, safely or otherwise. It is a standard based on toxicological data. The ADI, it must also be stressed, is not actually a scientific measure, but a regulatory construction based on scientific information. It combines an evaluation of toxicological data from animal studies with a "safety factor". It is as good a measure as the data it describes and the judgements made by those mandated to generate the numbers.

Either a temporary or full ADI , or a guideline level can be issued as a result of a JMPR assessment. The JMPR uses temporary ADIs and guideline levels to deal with pesticides for which it lacks sufficient data to issue an ADI. Thus, a chemical can move from an ADI to a temporary ADI, or even to a guideline level, if JMPR believes it lacks data for evaluation. The temporary designation does not reflect an assessment about the level of danger posed by a chemical, although it might seem that way to a layperson.

What is the status of these guideline levels and temporary ADIs in the Codex system? Until the past few years, the temporary ADI and guidelines were published along with the standards. This situation has changed. Now, once a Codex standard is adopted, the ADI – or the temporary ADI or the guideline level – is appended in a footnote. Codex has a standard (MRL) for captan, for example, but JMPR has given captan a temporary ADI.[21]

The ADI, or acceptable daily intake, is used in combination with other data, developed by the FAO panel in JMPR. These data incorporate information on the pesticide residue, its rate of deterioration in the natural environment, and something called "good agricultural practice". All three of these matters are assessed, and the FAO panel recommendation reflects the combination of these mainly qualitative assessments.

"Good agricultural practice" is the most difficult of these assessments to understand. "Good agricultural practice" is intended to reflect the practical capacity to reduce the amount of pesticide being used without effecting its usefulness as a pesticide. One Canadian official described "good agricultural practice" as meaning that "the agricultural business could not survive the loss of this compound at this time."

Good agricultural practice is a measure derived mainly from the supervised field tests run by the manufacturers to determine the minimum amount of pesticide that is required to achieve the desired effect. In different climatic conditions, the minimum amount of pesticide necessary "with good agricultural practice" to produce the desired effect will be different, so it is advisable to have field tests in more than one region of the world.

This occurs rarely, however. Supervised field tests are not conducted systematically in every region or country. Field tests are conducted only if the regulatory authority in a specific country requires them. Since many countries do not require registration of pesticides, or local field tests for pesticide registration, such tests are often not done. Moreover, field tests are costly. Countries are often reluctant to undertake them, and manufacturers protest any requirement for extensive field testing in several different countries. As a result, "good agricultural practice" refers to an extrapolation from whatever data is available from whatever field tests are done, in order to determine the minimum amount of pesticide that can be used effectively.

Good agricultural practice is considered an adequate measure, in spite of these limitations, because it usually indicates the minimum amount of pesticide that can be used under the best agricultural conditions to achieve the desired effect. This minimum sets a "standard" for the use of pesticides in less than optimal agricultural settings. No one suggests that good agricultural practice is the rule in every country that relies upon Codex standards, or that the minimum amounts of pesticides are used in each case. Good agricultural practice has a different significance. It is used, along with the ADI, as a way of determining the standard for the acceptable maximum pesticide residue on any crop.

The WHO expert panel in the JMPR is responsible for creating the ADI. The FAO panel arrives at the appropriate "good agricultural practice" and an assessment of the residue and its rate of deterioration in the natural environment and on food. These two assessments are then combined to create the JMPR recommendation for the MRL that is debated by the CCPR. Although the CCPR can, and does, alter the proposed MRL, it does not revise the ADI, but refers a pesticide back to the JMPR if problems are identified.

The JMPR can, and does, set standards for chemicals that some countries have defined as carcinogens. As the Secretary of the JMPR notes:

> The committee is free to give an ADI to the most potent carcinogen in the world. However if that is contrary to what was done in the past, they have to say why they did that. This is right because I cannot impose a code on them; I can only provide experience. Otherwise it would not be an expert committee.... It has to be based on the opinions of the experts.

What this means, in practice, is that the JMPR maintains no equivalent to the so-called cancer policy adopted in the United States, nor any adherence to pre-established rules about what will be acceptable as an ADI. If a "no effect level" can be observed in at least some toxicological tests, a chemical is a candidate for an ADI, and thus also for a standard or MRL. According to a JMPR official, if a "no effect level" cannot be observed:

> You have two alternatives, if it is avoidable or unavoidable. If it (the chemical) is unavoidable, the best thing is to achieve the lowest technical level. You have to be realistic. It would be very easy to pontificate and say a chemical should be banned. How do you do that? You find that the government will stop all the food industry from using plastic material (potentially vinyl chloride residues)?...Give us an alternative. You cannot force man back into a cave. So in the case of a residue that is unavoidable, we say the lowest technological feasible level is permitted. So then, some industry will supply the data on the lowest technological level feasible, with the best technology. However if you rely on that, you will penalize the developing country industry...(thus we depend upon ) having the experts to sort it out and (to decide) which model is best (for their assessment).

The decision about whether an ADI is appropriate is left to the JMPR.

What about the "no effect level" or thresholds that are used to arrive at an ADI? The JMPR dealt with this problem in its 1983 report, taking the position that no effect levels can be observed for most chemicals. In response to questions about the validity of a no effect level, the term "no effect level" was changed in 1983 to "no observed effect level". The JMPR report suggests that any method of risk calculation is subject to as many questions as the no-effect/safety factor approach has been.

TABLE 9

*National Acceptances of Codex Standards by Pesticide 1984*

| Pesticide | Number of pesticide food standards possible | Number of countries reporting | Full accept of various types | Limited accept of various types | Non-accept of various types | No Response |
|---|---|---|---|---|---|---|
| Aldrin | 775 | 31 | 44% | 6% | 12% | 29% |
| Bromophos | 372 | 12 | 26% | 24% | 17% | 2% |
| Carbaryl | 960 | 30 | 27% | 13% | 17% | 39% |
| Chlordimeform | 304 | 19 | 30% | 11% | 48% | 4% |
| Chlorobeniziliate | 162 | 27 | 30% | 11% | 19% | 33% |
| Cruformate | 34 | 17 | 30% | 18% | 15% | 6% |
| Diazinon | 748 | 17 | 30% | 6% | 15% | 37% |
| Diaxathion | 350 | 25 | 28% | 10% | 10% | 37% |
| Endosulfan | 114 | 19 | 40% | 14% | 32% | 1% |
| Fenchlofos | 256 | 32 | 15% | 13% | 1% | 43% |
| Folpet | 434 | 31 | 18% | 9% | 0.5% | 42.5% |
| Hydrogenphosphide | 320 | 32 | 40% | 3% | 5% | 35% |
| Methidathion | 444 | 12 | 35% | 21% | 13% | 0% |

## The Status of Codex Standards

In 1984, approximately 2500 food-pesticide standards were at some stage in the Codex system. 1600 of these standards had been officially adopted by Codex. Forty-two countries had replied to a request to accept these 1600 Codex standards as their national standards, although not all the replies were positive. The actual rate of acceptance of the Codex standards, and particularly of their full acceptance, is considerably lower than these figures would suggest. We examined the situation in 1983, the latest year for which complete data was available. Taking every fourth Codex standard, and calculating the percentage of its full and limited acceptances, and the proportion of nonacceptances, we arrived at Table 9.

TABLE 10

*Acceptances of Codex standards by country (1984)*
*(three regions only)*

| Country | Full acceptances | | | Limited acceptances | | Target acceptances | | | Non-acceptance | | | |
|---|---|---|---|---|---|---|---|---|---|---|---|---|
| | F | FT | FE | L | LT | T | TF | TL | N | NFD | NDC | NND |
| Canada | 288 | 148 | | | | | | | | 112 | 4 | 16 |
| USA | 17 | 141 | | | | | 2 | | | 201 | 55 | 405 |
| UK | | | | 2 | 25 | 345 | | 273 | | 199 | 4 | 16 |
| Greece | 146 | 50 | 17 | | | | | | 64 | 12 | | 38 |
| Netherlands | | | | | | | 135 | | | 10 | | 62 |
| Portugal | 4 | 35 | | | 580 | | | | | | | |
| Switzerland | 3 | 2 | | | | | | | | 111 | 43 | 13 |
| Spain | | | | | | | | | | 3 | | 42 |
| Cyprus | 213 | | 20 | | | | 609 | 1 | 8 | | | |
| Kuwait | 288 | 198 | 24 | | | | | | | | | |
| Israel | 3 | | | | | | 1 | 227 | | 16 | 17 | |
| Liberia | 276 | | | | | | | | | | | |
| Libya | 205 | | | | | | | | | | | |
| Tunesia | 179 | | | | | | 6 | 163 | 3 | | | |
| Jordan | 388 | | | | | | | | | | | |
| Iran | 62 | | | | | | | | | | | |
| Monaco | 4 | | | | | | | | | | | |
| Yemen | 29 | | | | | | | | | | | |

F and FT indicate full acceptance, probably accompanied by national legislation. FE indicates a standard for residues on foods that exist because of involuntary applications of pesticides (pesticide contamination, etc.). Limited acceptance, or L and LT, means that the country involved has a standard that is not less stringent than the Codex limit. In this case, the country undertakes not to hinder the importation of food that complies with the Codex standard. T, TF, an TL are national comitments to adopt a full or limited Codex standard after a specified period of time. Only NND represents total nonacceptance of the Codex standard. NND signifies that some conditions are applied to food with the residue distributed nationally. NFD means that the Codex standard has not been accepted but that foods meeting the Codex standard can be distributed freely within the country in question.

Relatively few of the approximately 120 Codex Member countries have accepted Codex standards, or even taken a position with respect to them. The rate of full acceptance of Codex standards is generally low, but nonacceptances are also less common than we might expect. Rather than formally register a nonacceptance, countries will take no position with respect to Codex standards. Codex standards exist for the highly controversial pesticides, such as aldrin, carbaryl, and diazinon. Yet, the rate of acceptance (or nonacceptance) for these controversial pesticides is no different from the other pesticides we surveyed.

We also examined the record of different countries with respect to the acceptances:

Data from African and Asian countries have not been included here for reasons of space. The pattern for their acceptances is similar to that illustrated in the table for the Middle Eastern countries. In general, the highest number of acceptances, and the least discrimination between different kinds of acceptance, come from the Middle Eastern countries. We believe that the Codex standards are most likely to be accepted by third world countries and used – inasmuch as any regulations exist – as national standards there.

A great deal of attention in CCPR and Codex is devoted to the problems of third world countries. Yet in 1985, the composition of the "ad hoc working group on pesticide residue problems in developing countries" was as follows: Of the total of 52 people in the room, 19 were delegates from industrialized countries, 7 were from WHO or FAO, 11 were from GIFAP, and 12 were from the developing countries.[22] Much of the attention to the problems of developing countries comes from the industrialized countries. The same situation applies to the JMPR. As a GIFAP participant noted: "now if you look at JMPR evaluations, you'll be impressed by the preponderance of data coming from North America and also Australia."

As of 1983, when these figures were made available, the United States was a reluctant participant in the Codex system, with a very high number of nonacceptances. In comparison, Canada was a very active Codex participant, even though Canada maintains a full regulatory system for pesticides and does not accept Codex standards without an independent national assessment of the data. The European countries participate in the Codex system, but appear to deal with each standard separately.

The differences between the European countries are noteworthy. Since the European Common Market also produces standards, a comparison between their standards and Codex standards is useful in interpreting the information here. Approximately 10% of the European Common Market and Codex standards were identical in the year of our survey. Less than half the products with a Codex standard moved freely within the Common Market. This suggests to us that the European Common Market standards have taken precedence over Codex standards for the Common Market countries, although we believe this situation is changing. Finally, we draw attention to the fact that even with these national differences, the level of acceptances of Codex standards is very low.

## Discussion

### *The Dual Mandate of Codex*

The Codex system is not the only standard setting or international body dealing with pesticides. The European Common Market has its own system of standards, with a process involving expert committees and national delegates. Several years ago, OECD issued some criteria documents, providing scientific assessments of specific chemicals and recommending tolerances. Since that time, OECD has withdrawn from the realm of formal chemical standard setting, but OECD still develops codes for such matters as good laboratory practice.[23] The International Association for Research on Cancer (IARC) issues scientific assessments of specific chemicals, including some pesticides. Finally, a relatively new United Nations organization, the International Program for Chemical Safety (IPCS), examines the impact of all chemicals, including pesticides, on health and the environment. IPCS has issued a number of its own criteria documents.

Of these organizations, we believe that only the European Common Market and Codex are very significant in international standard setting. This is true, even if Codex standards are not widely accepted by its Member countries. Compared to others, both Codex and the European Common Market have a fully operational process for arriving at standards, and a comprehensive method of conducting chemical assessments. These organizations are well established. As is the case with ACGIH, the voluntary status of their standards disguises their importance.

The contrast between both Codex and European Common Market, on one hand, and IARC and IPCS, on the other, is important for the study of mandated science. Those who seek a comprehensive and neutral scientific assessment might well wish that Codex was more like IARC, or even that the JMPR used the same criteria as IARC in its assessments. IARC relies upon the publicly available scientific literature exclusively, and is not beholden to industry for access to its data. It uses a conventional peer review process to evaluate research. It is able, on occasion, to replicate a study or to conduct further research to answer a difficult question. It sets its own timetable for the publication of research and publishes all its research.

Nonetheless, we would argue that the IARC will never supplant either the JMPR or the European Common Market with respect to scientific assessment. Nor have we seen evidence that IARC reports are relied upon in any significant measure by the JMPR. The reason lies in the nature of the data that must be used when industrial interests – and in particular pesticides – are under observation. IARC cannot rely upon the companies' disclosure of the information that is needed to conduct its assessments, because IARC does not respect the confidentiality of proprietary data. Since IARC monographs are not designed to be immediately useful to regulators, and IARC does not weigh regulatory considerations in its assessments of particular chemicals, IARC has no lever to use in gaining access to information. Finally, IARC has only resources for a limited number of assessments, too few to provide the basis for a comprehensive system of standards.

IPCS deals with all chemicals, not simply pesticide residues on food. As such, it is able to take a broad health and environmental perspective, and to consider industrial pollutants as well as pesticides. IPCS reports to WHO, and is thus responsible to a single mandate, the protection of human health. IPCS also brings together different UN agencies and other organizations conducting chemical assessments, to share information. Again, a student of mandated science might wish Codex was more like IPCS.

We believe that IPCS also will not replace Codex or the JMPR. IPCS is caught in the maze of the United Nations and other organizations that make up its membership. Their conflicts become conflicts within IPCS. IPCS made an effort to secure an independent source of data for its assessments. In this, we have been told that IPCS was only marginally successful. Only JMPR has the power to convince industry to release the data, because only JMPR ties this information directly to the standard setting process. In practice, IPCS now draws very heavily upon the assessments of JMPR with respect to pesticides, and its influence is, as yet, very limited.

Why are the standard setting activities of Codex and the European Common Market so significant? Our answer is that the explicit connection of health and safety with trade and national interests contribute to the dominance of Codex and the ECC in standard setting. Codex and the European Common Market are able to set standards, because their member participants can see the direct relevance of standard setting activity to their own national and economic interests. Their members can deal relatively easily with health and safety issues, because in doing so, they also address trade questions. As trade organizations, both Codex and the European Common Market have the necessary power over their members – particularly in the case of the European Common Market – to ensure at least some level of their active participation and co-operation.

### *Standards and Trade*

To say that standards and trade are intimately related is to say no more than do the official publications of Codex. It is important to understand the intricacies of the relationship between standards and trade. Different standard setting organizations create different environments for trade, and in doing so, privilege some participants and engender different trade relationships.

Codex standards are primarily connected to export decisions. A Canadian official described the situation to us, as follows:

> There are pressures from particular countries who are using a chemical – aware that it is leaving a residue in the food that they may be exporting and they want to protect their exports.

Food exporting countries cannot send their products to countries if they cannot meet the local standard for pesticide residues. Oranges with too much residue, or sprayed with a pesticide not registered for use in the United States, cannot be exported from Brazil or Israel to the United States. The more dependent a country is upon one or a few food crops as a major source of its export revenues, the greater will be its concern for standards in the countries to which its food is being exported.

Countries that export food crops care about the standards in the countries where they sell their products. Countries that import food crops may use standards as a non-tariff barrier to trade. The latter is never discussed openly, but is widely acknowledged in informal discussions in standard setting organizations. As noted in the introduction of this book, standards function something like the gates in a pinball machine. They direct the flow of trade, but are often not themselves the purpose of the exercise. As one GIFAP observer commented to us frankly: "There are many countries who use the MRL as a political instrument..."

As might be expected, standards have quite a different significance in different contexts. The best way to illustrate this is by taking four hypothetical examples of export trade relationships and showing how they would be influenced by the acceptance of Codex standards. Decisions about standards have a different significance when seen from the perspective of an underdeveloped or industrialized country. They have a different significance for the multinational pesticide manufacturers than for local farmers.

Our first example is the hypothetical case of a developing country, in which agriculture is a major source of economic development. Let us assume that national standards are nonexistent in this hypothetical country, and the the resources for local research to develop and implement standards are meagre.

In this case, accepting Codex standards has double-edged consequences. On one hand, without standards, our hypothetical country is unlikely to be able to export its foods to many countries. Accepting Codex standards ensures that food exports *can* move in trade. Without standards as well, food is likely to be considerably less safe. Codex standards also take the place of a national system of standards. On the other hand, accepting Codex standards places a serious burden upon the domestic producers. They have to meet the same requirements as importers for the acceptable levels of pesticide residues on their food. This increases the costs of their agricultural production, and reduces their freedom to use pesticides to their maximum advantage. On balance, if food shortages are common and domestic food production predominates, our hypothetical country would be reasonable if it ignored the Codex standards, and failed to generate any national standards to replace them.

In light of our hypothetical example, it is interesting to note the comments of a GIFAP participant at the Codex meetings. He compared the situation in developing countries with that of the industrial countries:

> The agricultural structure is more developed in cases like continental western Europe than it is in some of the third world countries, (especially since the latter are operating under) hot conditions. Therefore it is possible for The Netherlands, to take an example, to say that (they) will have a post harvest interval of 70 days...because they can arrange it. They'll pick their cucumber crops if they are ready, and then they'll do nothing for several more days...They can organize themselves in that way. This is not the case in the third world.

In order to meet the Codex standards, crops cannot be sprayed too often, even if climatic conditions produce an outburst of insects or fungus. They cannot be harvested – or must be stored after harvest – until the interval between spraying and sale required by the standard has elapsed. They cannot be exported unless the pesticide residues have degraded. An industrial country has the resources to plan its spraying and harvesting to ensure compliance with the Codex standard. Most third world countries do not.

Our second hypothetical example is a third world country in which multinational food producers play a major role. For such a country, and the multinational companies operating in them, the benefits to be gained by the access to export markets for food crops (on the basis of standards) outweigh the increased costs of producing food that conforms to Codex standards. Because of their size, multinational companies are usually geared to production methods that can accommodate the Codex standards. Our research suggests that many third world countries do accept Codex standards.

FAO officials, Codex officials and CCPR participants all stress the importance of third world countries accepting the Codex standards. These officials indicate to us that the third world is slow to accept Codex standards, in spite of our evidence to the contrary. Our sense is that these officials have a "mission" with respect to third world countries. Their mission is to get third world countries to accept Codex standards (even if industrial countries do not), and their portrayal of the lagging acceptance rate in the third world is useful in accomplishing this mission.

An FAO official summed up an attitude we heard expressed often:

> We often hear: "you persuade the US to accept the limits, and we'll accept the limits" from the underdeveloped countries. We get complaints from the underdeveloped countries. We say to them:" If you don't have a regulation, you should say that you accept the limit, or if the limit is higher (than you want), choose the limited acceptance route". The justification is the fact that *all* countries need to import part of their food supply. And the purpose of a standard is to facilitate trade.

The introduction of "limited acceptance" was considered a "breakthrough" in the eyes of at least one FAO official, for it meant that food can be exported, even if Codex standards were not used for domestic agricultural production. "Limited acceptance" of Codex standards serves the purpose of export trade, but frees domestic producers from some of the domestic constraints that Codex standards impose. Unfortunately for domestic producers, limited acceptance of Codex standards also results in the lack of domestic barriers to the importation of food. As the FAO official said:

> Limited acceptance means you are prepared to accept the maximum limit for pesticides (the Codex standard) even though a pesticide is banned in your own country, but you permit imported products on the grounds that you need the trade.

Our third hypothetical example is an industrial country with a full complement of national standards. Let us assume that these national standards are more stringent than either Codex standards or the standards adopted in many other industrialized countries. Our hypothetical country rejects the idea of active participation in Codex, and accepts none of the Codex standards. The stringent national standards protect local agricultural producers from competition from imported products, acting as a non-tariff barrier to trade. The stringency of national standards also ensures that food products being exported will be acceptable in any other country.

Stringent national standards are conducive to the practice of "dumping", the export of foods with excessive pesticide residues or of restricted pesticides to a country with less stringent standards. The stringency of national standards places constraints upon domestic agricultural production. Deciding not to accept Codex standards removes the necessity of adhering to their particular limitations. A Canadian official told us that there was a "whiff of truth" in the claim that standards are associated with "dumping". He went on to justify the practice, which makes us think that it is more routine than is usually acknowledged. He said:

> But the question is then what is the need of the country in which they are dumped. That is something people don't often think about.

The reader might be excused for thinking that our third hypothetical example refers to the United States, but the United States – while it does not accept many Codex standards yet – has very recently become an active participant in Codex. American government officials deny that standards are used as non-tariff trade barriers. One such official stated:

> We don't use pesticide regulations or tolerances for trade purposes. I know it's done, but we have a lot of other mechanisms that would look out for the Florida (the example being used in our discussion) commercial interests. We could put a quota on them. I've never seen anything that would indicate that the US is playing that kind of game.

We cannot assess the import of his denial. The American attitude towards Codex was a positive one when the interview was conducted. We also cannot imagine an interview with any government official in which the use of standards as a protectionist measure would be discussed openly. There are a number of other countries that might be described by this hypothetical example.

Our final hypothetical example is an industrialized country that supports the Codex system, and accepts many of its standards. Acceptance of Codex standards increases the predictability of export trade, ensuring that the "rules of the game" are known in advance. The fact that Codex standards are less stringent than most national standards is also important. To match the Codex standards and promote fair competition, domestic farmers will argue that the stringency of national standards should be reduced. Accepting Codex assists those lobbying for less stringent national standards by providing an alternative to stringency, and associating this alternative with international co-operation.

Pesticide manufacturers support the acceptance of Codex standards for the simple reason that such standards are a substitute for national registration in many countries. If Codex standards can be accepted, then international manufacturers will only be required to participate in a single system of assessment, the Codex system. It is worth noting that it is GIFAP, an international organization of pesticide manufacturers, that is most active in CCPR. Indeed, GIFAP plays a very helpful role in Codex. GIFAP publishes, at its own expense, several useful guides on safe handling of pesticides for Codex. When asked how critical it is that the governments climb on board the Codex system, a GIFAP member replied:

> Well, it would make our life an awful lot easier on Jupiter! If there were internationally acceptable MRLs, then we wouldn't have to spend so much registration time and money on so many dozens, if not hundreds, of different crop pesticide combinations for different countries. It would save an awful lot of time for us...

The use of these hypothetical examples is necessary because the actual reasons for support or rejection of Codex standards are not available to us. They are connected with decisions about national security and export trade. However, we were told several anecdotes about countries that redirected their food shipments – literally en route – because the residues did not meet the standards in the country to which the food was originally directed.

### *Science and the Uses of Language*

Both CCPR and JMPR use the term "acceptable daily intake" to refer to scientific information contained in their standards. They use the phrase "acceptable daily intake" in a technical sense, and in a manner that is not intended to convey how much of a pesticide is actually consumed by any individual or within any national context.

The JMPR does seek information on actual pesticide intakes, but this information is not available for new pesticides, and is often suspect for pesticides currently in use. The JMPR also suggests a method for extrapolating how much of a pesticide is actually ingested, either from food consumption patterns (this is called a "theoretical daily intake") or from whatever marketbasket surveys are available (called, "estimated daily intake"). But neither the JMPR nor CCPR themselves use these methods to set standards. The JMPR suggestions are for the countries that accept Codex standards as their own. The JMPR itself relies on toxicological data.

Similarly, "good agricultural practice" is not an assessment of how pesticides are actually used by farmers in any country. Good agricultural practices will vary in different countries, with respect to different climatic and soil conditions. They will vary according to the technologies and resources available to farmers and the cultural patterns of agricultural production. JMPR members know this, but the measurement of good agricultural practice is related to the optimal performance of a pesticide irrespective of its location.

In the same manner, the "temporary ADI" is a term with different connotations in the JMPR and in everyday use. In the JMPR it does not reflect the level of danger of a pesticide, but is instead connected to the data available for its assessment. In everyday language, the shift from an ADI to a temporary ADI seems to imply that a new danger has been identified for the pesticide in question. There are even inconsistencies in the use of such key terms as "pesticide residues". If one examines the national regulatory practices, in some countries, the term "pesticide residue" refers to the residue on the whole product, and in others, only to the residues on the edible portion of the food.

Understanding the mixture of natural and specialized language is critical to understanding all of these terms as they are used in Codex. The terms "good agricultural practice" and "acceptable daily intake" are in fact operational terms that have a specific meaning in the context of Codex methodology. They have quite a different – indeed occasionally contradictory meaning – in everyday language.

We witnessed the confusion about language in a discussion of the report of the committee on regulatory practices in the 1985 CCPR assembly, where one third world delegate argued that the ADI should be responsive to local conditions. We know, as do members of the CCPR, that toxicological testing is conducted with animals in laboratory settings, and thus the jurisdiction in which the tests are conducted is irrelevant. Yet here was an accredited delegate from a large country requesting that the ADI research accurately reflect the local food consumption patterns and local conditions of his country. His questions were not answered with ease.

The problem stems from the fact that the term "ADI" does not refer to a natural phenomenon, but to a series of judgements made as a result of a regulatory procedure. These judgements incorporate knowledge of a scientific literature, and are scientifically informed, but many of them are not scientific in a narrow sense of the term.

The judgements involve many things. They involve an evaluation of the choices made in the design of the relevant toxicological research – judgements concerning the use of particular species, for example. They involve an extrapolation from animal studies to human populations, an extrapolation from what is observed to what is not. The ADI reflects the application of a safety factor, a measurement that is almost exclusively judgemental in origin. It is combined with an equally judgemental assessment of the residues and of "good agricultural practice". In summary, even if economic considerations played no role in JMPR decisions, a number of other regulatory judgements do.

As well as having to make a series of regulatory judgements, the JMPR operates under a particularly influential set of constraints, none of which is typical of science. This point is important for the study of mandated science. The JMPR conceives of itself as a scientific group, but the JMPR members only evaluate the research of others. They cannot normally replicate it to resolve their doubts.[24] The data they evaluate are a product of political conditions and interests that have little connection with science. The time frame for assessment is short, and JMPR decisions must be rendered conclusively if at all possible. These decisions have immediate political and regulatory consequences, a fact that must impinge upon the consciousness of the JMPR members to some degree. The members themselves are often regulators, acting in a different capacity during the ten day period of the JMPR meetings. As regulators, they have an on-going, and only marginally scientific relationship with the industries that produce and use the pesticides they evaluate. Finally, the JMPR usually does not set its own priorities, but responds to the requests of another body, whose mandate is only partly scientific.

The operationalizations conducted to arrive at an ADI or "good agricultural practice" are necessary because of the constraints imposed upon the JMPR members. They incorporate the personal, political and scientific judgements made by the scientifically trained members of the JMPR into a single regulatory measure that is useful to set standards. By using technical terms and operational definitions that have limited reference to their everyday meaning, intentionally or not, Codex is able to mask the conflicts inherent in its organization and approach to standard setting. The final product, the ADI, seems more scientific than it is, although it does incorporate scientific judgements as well as nonscientific ones. The JMPR's final assessment is the "science" that is taken into account by the CCPR and Codex.

Insiders, and those fully aware of the intricacies of the Codex system, are not likely to be misled by the inconsistencies in Codex, JMPR, and in the various national practices. They have sufficient background knowledge to fill in the missing information and interpret statements about pesticide residues accordingly. Insiders understand these limitations of the JMPR assessments. For them, the operationalizations are a necessary and inevitable part of standard setting. Insiders can also gauge when to rely upon JMPR assessments, and when these assessments should be tempered by other knowledge.

Someone lacking this intimate knowledge of the Codex process – and this includes members of some of the delegations – is unlikely to develop such a finely tuned reading of the proceedings. The CCPR delegate who raised questions about the ADI was not, in spite of his delegate status, an insider. It was with some difficulty that the other delegates tried to explain the process of standard setting to him. For him and others, the technical language masks the uncertainties that are felt even by members of the JMPR. Confusion between everyday terms and operational technical ones lends a solidity and stability to a process of assessment that is invariably less scientifically certain than its participants would sometimes like to claim.

We believe that an insider-outsider relationship has existed within CCPR, among the various national delegations for some time. The clash of technical and everyday language apparent in the interchange between the delegate who failed to understand the nature of the ADI, and those who responded to him from CCPR, is just one example of a continuing problem. The length of time it has taken CCPR to arrive at an agreed upon description of its process of assessment provides further evidence for our contention. Codifying the procedures for arriving at an ADI has been difficult, we suggest, because it highlights the differences among members in their understanding of the CCPR process.

The insider-outsider difference in both the JMPR and CCPR is important for another reason. We have referred to the fact that JMPR members, who are often regulators in another context, are not supposed to know about data that has not been officially submitted to the JMPR, nor are members supposed to take their national interests into account. For an insider, well-versed in the constraints imposed on the JMPR, this situation seems credible. Insiders are also attuned to the procedures that are used to control some aspects of national bias. To an outsider, the situation of JMPR members is untenable. To an outsider, it seems unlikely that a member of the JMPR could conduct assessments without taking account of national interests or of the regulatory information available in another context.

Finally, the insider-outsider difference has quite a different significance. If one assumes that decisions about standards should be debated in a public forum and that laypeople should participate in them, then the difference between the insiders' language of technical terms, and the outsiders' understanding of those same terms is an important one. It is worth remembering that terms such as ADI and TLV are used in the popular press, and are interpreted by prospective participants in the public debate to provide information about public safety.

The layperson's interpretation is likely to be inconsistent with the assessments of the JMPR. The layperson is likely to view the ADI as exclusively scientific in origin. He or she is likely to think that an "acceptable daily intake" refers to the amount of a pesticide is actually ingested. He or she is likely to think that "good agricultural practice" refers to how much pesticide is, and should be used in each country. The laypersons' understanding of the common terms constitutes a barrier to informed participation in decisions about international standards. For those who discover the differences between the JMPR's use of terms and their own, the result is likely to be disillusionment with protection that is offered by Codex or other standards.

Members of the JMPR are adamant that their assessments are relatively free from bias. In this, the members of the JMPR are similar to their counterparts in ACGIH. They regard themselves as "watchdogs" of industry and as advocates of public health and safety. They respond to criticisms by pointing out the various measures they take to deal with the constraints upon their assessment, the manner in which they negotiate with industry for the release of data, and the difficulty of arriving at any better means of setting standards. They do not believe that a more public process or a new organization would solve problems which are endemic to standard setting. We will return to their response in chapter seven.

Chapter Five

# Political Chemicals
# Case Study Three: The Toronto Lead Controversy

*with William Leiss*

Lead is known to cause harmful effects at certain levels of concentration in the blood and organs, but the standards for lead are controversial everywhere. Why is this so? If the toxicity of lead is not really in dispute, why are so many governmental agencies involved? Why are so many commissions, committees and inquiries necessary? What is forcing the growing stringency of the lead standards, and what scientific issues keep the debate going? Or is science being debated?

In Canada, the most important lead controversy occurred in Toronto between the years 1970 and 1979. It concerned the airborne emissions of lead from a number of metal smelting operations. The companies extracted the lead from electric batteries for recycling. The complaint was that "fugitive" emissions from the smelting operations and pollution from accumulated industrial debris – piles of scrap metal – had contributed to very high levels of lead in the dustfall and soil in the residential areas adjacent to the smelting companies.

If logic were the only determinant, the piles of scrap metal should have been cleaned up immediately and pollution control equipment installed. Instead, actions were taken that were most unusual from a Canadian perspective. The regulator issued its first "stop order", only to have it overturned. The case went to court, not once but several times and with respect to several issues. An environmental assessment board hearing was the last stage in a series of hearings and committees, but its report only fuelled the controversy. Three levels of government were involved, two of them as opposing parties in the dispute. It took a long time, some years in fact, before the complaints were resolved.

The most contentious lead standard in the Toronto controversy was "worked out on the back of an envelope". In fact, some standards originated with ACGIH, and some were extrapolated from the TLVs and applied to environmental pollutants. Yet the whole debate was seemingly about standards. Three matters were at issue: the proof of a hazard, the relative stringency of existing and proposed standards, and the willingness of government to enforce them. Scientists were involved in the controversy as government consultants and committee members, as expert witnesses and as advocates. The debate had the appearance of being about the scientific basis of the lead standards, but only rarely was it a scientific debate.

The lead case is important to the study of mandated science for a number of reasons. It provides an opportunity to observe a public controversy about standards. Because scientists are involved in quite different capacities, it illustrates some dimensions of the relationship between science and public policy. The use of lawyers and courtroom tactics is quite unusual in the Canadian context. For this reason, the relationship between legal and scientific norms is brought into sharp relief by the study of the Toronto lead controversy. Finally, in the course of our study, we coined the term "political chemical". A political chemical is one that has sparked a political debate about the broad issues of chemical or pollution control. A political chemical is a lodestone for the public debate about standards. Lead was, and is today, a political chemical.

## Background Information

At the time of the Toronto lead controversy, Canada accounted for about sixteen percent of the world production of lead, one third of which was produced by secondary lead smelting.[1] Most of the secondary lead smelting was done in Toronto by three plants, two of which were active in the Toronto lead controversy. It is important to note that lead emissions do not come solely from either the secondary production of lead or from lead processing.[2] For example, about seventy percent of the lead emissions in Ontario at the time of the controversy was due to lead in gasoline. Other metal processing industries constitute major sources of lead emissions, as does waste disposal. Today the debate about the lead standards focusses on lead in gasoline and lead from other industrial sources. These were minor issues in the Toronto lead controversy.

Secondary lead smelting was relatively profitable during the early phases of the Toronto lead controversy. At least one of the companies also experienced a significant surge in profits in the mid- and late 1970s. As a result, the lead companies were in a position to allocate resources to pollution control. Profit levels in the industry have fallen since 1980, and today, the prospects for secondary smelting companies are poor. Some improvements have been made in the lead smelting operations in the last few years. Although citizens still complain that the plants are unsightly and that the plants cause noise, dust and pollution problems, the smelting operations are no longer at the centre of a public debate about pollution.

At its height, the Toronto lead controversy involved many different groups. From the public side, two citizen groups, the provincial labour federation, a political organization, two political parties, and an advocate organization all took active roles. The advocate group was made up of lawyers, who were employed on a staff basis to assist in legal advocacy on environmental issues. From the government's side, no single agency had regulatory responsibility. The local

Board of Health, the City Council, several provincial Ministries and at least three federal Ministries were all involved. The scientists who participated did so mainly as expert witnesses, on one side or the other of the conflict. Finally, industry was composed of four quite different companies: a small family operation, a subsidiary of one of the largest Canadian companies, and two companies owned by American parent firms.

Jurisdiction over lead pollution is a complex matter reflecting the relationships among the various levels of government and their departments. The federal government departments were not very active in the actual controversy. In fact we were told by one federal official that the Toronto lead controversy was regarded as "a local situation." Yet the federal department, Environment Canada, conducted assessments and produced reports. It was empowered to set standards, although the federal *Clean Air Act* was designed mainly to assist provinces develop air quality standards that could be enforced under provincial legislation.

"National Ambient Air Quality Objectives", that is, federal air quality standards, were set by federal-provincial expert committees. These standards were not regulatory, since there was no provision for an "offence" in the law. A federal official described the *Clean Air Act* in the following terms: " (it was) not a mandatory sort of thing, but enabling legislation." The federal Cabinet also had the power to prescribe national emission standards from particular stationary sources, if these emissions constituted a significant danger to health or resulted in violations to international treaties.

The method for arriving at the air quality objectives was an informal one, and it involved extensive industry consultation. It is worth quoting a description of the process by which the early standards were set for lead in gasoline, because there is no evidence that the process for the air quality lead emission standard was any different:

> So then the basic process for that (lead in gasoline, in his example) regulation was discussions between the environmental people in the Environmental Protection Service and the industry, that is the automotive industry and the lead additive industry. And at the very senior level (there was) a negotiation of what level of lead everyone would tolerate in gasoline. It was done by direct negotiation, much in the same way that a labour negotiation comes about. Each side picked a starting point, and each side negotiated to something they could live with. I think that is the simplest way to put it.
>
> In those days there was no public input. I shouldn't say this but I will. The evidence that Health and Welfare used to substantiate the regulation was as flimsy as the paper it was written on. They based it on some studies that were ongoing, and what was done in the United States.

A number of different federal departments are involved, directly or indirectly with the federal standards for lead. These include: Environment Canada, Health and Welfare, Consumer and Corporate Affairs, Labour, Transport and two quasi-governmental advisory bodies, the National Research Council and the Science Council. Negotiations among them with respect to jurisdiction and responsibility are a critical part of the picture of standard setting in Canada.[3] The science advisory councils published reports that took contrary positions on the dangers posed by lead and the adequacy of the standards. Neither Council is integrated into the regulation of chemical hazards, and their assessments or reports do not constitute the basis for regulatory decisions.

The most active governments in the Toronto lead controversy were the municipal council and the provincial authorities. The municipality was the original regulator of hazards such as lead pollution, and it had some rudimentary standards. The Toronto Board of Health, acting under the mandate of the provincial *Public Health Act,* still had the power to set standards and issue abatement orders, but the provisions for their enforcement were unclear.

In 1967, the provincial *Air Pollution Control Act* was amended, and the Air Management Branch under the newly created provincial Ministry of Environment took over the primary responsibility for setting environmental standards and enforcing them. The Ministry of Environment used interdepartmental expert committees to set its standards, thus limiting public involvement in environmental standard setting. A provincial *Environmental Assessment Act* was passed in 1975, creating the first formal process of environmental assessment at the provincial level.

The shortcomings of the provincial environmental legislation became an important element in the Toronto lead controversy. Under the *Environmental Protection Act,* the Ministry was empowered to act only if there was an immediate danger to human life, the health of persons or to property, and the Ministry of Environment had considerable discretion about taking court actions. Under the *Environmental Assessment Act,* the Minister also had discretion about whether to call public hearings. The environmental laws fuelled the controversy. It was difficult to prove immediate danger in the case of the lead pollution, and the opportunities for court action and public hearings were constricted.

The provincial Ministry of Environment had two courses of action with respect to pollution offences. First, it issued control orders, outlining a formal agreement between the company and the government about the steps to be taken to clean up the pollution. The terms of the control orders were negotiated with the pollutor, and they were designed to impose a schedule for pollution abatement. If the abatement schedule was being followed, the issuance of a control order exempted companies from prosecution for pollution related offences while the order was in force. Second, the Ministry issued stop orders, in this case prohibiting the industrial activity that was creating a pollution problem. Faced with violations of either a control or a stop order, the Ministry then initiated discussions with the offending company or took the matter to the courts.

The legislation under which the Ministry operated gave the Ministry the power to issue permits for the emission of specific levels of contamination.

These permits were called certificates of approval and program approvals. These approvals were individual contract agreements between the Ministry of Environment and the firm or municipality proposing to emit the pollutants.

In fact, in Ontario as elsewhere, the Ministry was forced to rely upon industry to do much of the basic monitoring and testing of emissions. The Ministry and its district offices monitored the implementation of its program approvals for individual companies, but it did so systematically only when new equipment was being installed and before the industry's operations commenced. The Regional and District Offices of the Ministry conducted periodic audits of the information submitted by industry. The Ministry lacked the resources for routine monitoring.

We have heard it said that the Ministry's inspections were highly reactive and that the "control orders" issued by the Ministry with respect to infringements of its program approvals were regarded by the Ministry as "regulatory failures."[4] We know that the Ministry faced a situation of declining resources in the period from 1978 through 1980, during which time the Ministry's activities to secure compliance with its orders suffered. Another analyst of this period has suggested that "(t)he inspection picture which emerges is one of an unplanned and often unsystematic operation."[5]

The provincial Ministry of Labour was also involved in the Toronto lead controversy. This Ministry was empowered to develop standards for airborne contaminants in the workplace, but in fact its standards at the time were simply guidelines. During the height of the Toronto Lead controversy, conflicts between the Ministries of Labour and Environment were inevitable, given the lack of procedures in either Ministry for standard setting and the unclear demarcation between occupational and environmental hazards.

The various provincial Ministers of Health who held tenure during the lead controversy refused to become involved. The situation with respect to the involvement of the Minister of Health was complicated, however, by disputes concerning the jurisdiction of the various governments involved in health care. For example, at the time of the lead controversy, an unrelated conflict existed concerning the proposed consolidation of several municipal boards of health into a single area-wide board of health. This conflict undoubtedly influenced the Ministry of Health's attitude towards the activist role taken in the lead controversy by one of the local boards, the Toronto Board of Health. A second conflict was between the Ministry of Health and the Ministry of Labour, both of which could legitimately claim (or refuse to acknowledge) responsibility for the health of workers (who were undoubtedly exposed to more lead pollution than the irate citizen groups). This second conflict was only resolved in 1978, with new legislation that formally delegated all occupational health matters to the Ministry of Labour.

## Standards in the Toronto Lead Controversy

The people we interviewed from the Provincial Ministries did not know the origin of their standards. However, in a paper issued in 1975, a Ministry of Health consultant described how the air quality standards were originally set:

> The air quality criteria was derived by taking one-tenth of the industrial occupational *threshold limit value* (the term is copyrighted) of 150 microgrammes per cubic metre. This standard for occupational exposure is based on only 8 hours exposure per day for a five day week. In order to meet the 15 microgrammes per cubic metre per 24 hour criterion, a design standard of 20 microgrammes per cubic metre for 30 minutes was set. This design standard was used to assess new and existing sources of lead emissions and was a standard determining, by use of the Pasquill-Gifford diffusion equation, the allowable emission from a source. If this 30 minute standard was met, the ambient air quality criterion of 15 microgrammes per cubic metre per 24 hours would also be met.[6]

In a different report, issued in 1973 by the Ministry of Health, the basis for the lead standards and for their proposed revisions was provided.[7] Extrapolations, involving several equations and estimates from various routes of exposure, were made to convert airborne standards – certainly ACGIH standards – to standards for dustfall, soil and blood leads.[8]

The standards used in the Toronto lead controversy seem to have originated with ACGIH, then, although it would have been impossible to determine this information from the debate itself. A second source of standards was the actions of agencies in other countries. For example, the 1973 report referred to the decisions made by the Environmental Protection Agency in the United States. This report contained a table of standards for lead in various jurisdictions.

The federal Department of Environment set a lead standard for secondary lead smelting in 1976. We believe that the ACGIH TLV was used. A report, written in 1974 by Departmental officials, provides the evidence for our contention.[9] It details the assumptions for the extrapolation of blood lead levels "from the TLV". The federal Department of Labour also set standards for lead. The Department of Labour relied on the American health and safety agency, OSHA, for its standards, which means, of course, that they used many standards that were identical to the TLVs. It also used assessments provided by Health and Welfare for an assessment of the scientific materials. It has been impossible to determine whether Health and Welfare used ACGIH or OSHA standards or NIOSH documentation in its assessments. It can only be said that, at the time of the Toronto lead controversy, the resources in Health and Welfare for the assessment of chemical hazards were limited.

The federal lead standard for secondary smelters set by the federal Department of the Environment was an emission, or prescriptive standard. Pollution was to be measured when it was released into the surrounding environment. In the Toronto lead controversy, advocate groups claimed that the lead smelting companies could conform to the federal emission standard by simply reducing the ratio of contaminants to air in their emissions. The provincial standard was a performance standard. In other words, measurements were to be taken at the point of impingement – or exposure – of the contaminant and on the public or the environment. The relative merit of prescriptive and performance standards was debated extensively in the Toronto lead controversy.

The federal lead standards were "objectives", or guidelines, The provincial lead standards could be either guidelines or regulatory standards. The categories used by the provincial Ministry of Environment were: regulatory standards (called "standards"), tentative standards (which could be used for abatement action, even if they had not been formally approved), guidelines and provisional guidelines. In other words, the provincial government had considerable discretion about the legal status of its standards.

The federal government prepared an internal report on lead emissions and standards in 1974, and a federal lead standard for emissions from secondary lead smelting was issued in 1976. The provincial standards for lead were originally adopted in 1968, when the provincial government acquired jurisdiction from the municipalities. Recommendations were made for revisions to the lead standard in 1972, but they were not adopted immediately. A report was prepared by a provincial Ministry official on the lead standard in December 1973. More stringent lead standards were then adopted in 1974. These revisions occurred at the height of the controversy, and before new standard setting procedures had been adopted by the Ministry of Environment. We think they were a result of the controversy, and not just of a review by Ministry officials.

Later, in 1978, the provincial Ministry of Environment outlined a formal standard setting procedure, claiming that this procedure had been evolving in the Ministry in the previous few years.[10] Scientific data and technical data were to be reviewed by a committee of experts from the Ministry of Environment and of Health, and the standards were to be approved by Cabinet.[11] Even with the new procedure, however, only some standards were to be developed with a full Ministry review.[12] For others, and when no other data were available for the Ministry's review, occupational data – and thus the TLVs – were still to be used as a basis for environmental standards.

## The Toronto Lead Controversy (I) – Early History

In 1966, nine children were admitted with lead poisoning to the children's hospital, and several hospital doctors began an investigation. The doctors located the source of the lead in the clothing of a worker at a local lead smelter, but one doctor continued testing for lead exposures on his own initiative. Then, in 1969, a complaint was made by a local resident to the Toronto Board of Health about lead pollution from the Toronto Refiners and Smelters Company. Another formal complaint (whose origin we cannot locate) was lodged with the Air Management Branch of the Ministry of Energy Resources – later to become part of a new Ministry of Environment – initiating research and discussions about lead pollution with the Canada Metal Company. These two complaints about two different companies set the Toronto lead controversy in motion. In 1970, in an unrelated instance, residents living near the Toronto Refiners and Smelters Company submitted a petition to the Toronto city government about noise and dust from the plant. Seventy people signed the petition, including the original complainant.

The city government had already been apprised of a possible problem with lead pollution from the smelters, because one of the city councillors was in contact with the doctor at the children's hospital, who was beginning to suspect that the source of lead poisoning was the smelting operations. The city council passed the citizens' petition to its Commissioner of Buildings, who had authority to investigate industrial nuisances. This put the complaint into the hands of the city's building committee. The city officials also requested that the Commissioner's staff, the provincial Medical Officer of Health and the newly created provincial Ministry of Environment investigate. In late 1972, however, the residents who had signed the original petition discovered that the city's building committee had not investigated their complaint.

They contacted a lawyer from an environmental organization.[13] The environmental lawyer determined that Toronto Refiners and Smelters had not obtained a permit for its battery crushing operations. The lawyer proposed that the citizens approach the Toronto Board of Health, and as a result, one of the citizens, the lawyer from the environmental group and a city alderman living in the affected area (and who was an environmental activist himself) began attending the meetings of the Toronto Board of Health.

The environmental law group also threatened legal action against the Ministry of Environment, with the original citizen complainant acting as the plaintiff. The proposed action concerned many issues, but the lawyer's letter noted the smelting company's lack of a proper permit and the Ministry's lack of action with respect to it.[14] The Ministry was given several days to respond to the threat, and eventually it did. In July 1972, the Ministry closed one of the battery crushing operations, and issued a stop order to Toronto Refiners and Smelters. This was the second stop order ever issued by the Ministry of Environment; the first stop order (we have not been able to identify the company involved) had been issued four weeks earlier and removed shortly after it was issued.

This second stop order remained in place until July 1973, when the Ministry indicated that the equipment installed by Toronto Refiners and Smelters was sufficient. A program approval was issued to Toronto Refiners and Smelters in January 1973, which specified the installation of emission controls as part of the permit. The conditions were fulfilled for the most part by June 1974, but in January 1974 the company was charged on another unrelated pollution offence. The citizens claimed, however, that the equipment, which was installed in response to both the program approval and stop order, failed to abate the dust pollution from the smelting operation.

We believe that all the legal threats and actions had another effect. The Toronto Refiners and Smelters Company, unlike Canada Metal, is a small family owned company, operated by a father and his two sons. The father was a victim of Nazi persecution in Germany, and since he came to Canada, he has been a community leader and philanthropist. The family felt that, under the circumstances, the father should be protected against any legal action. What might be, in any other circumstances – and indeed was, in the case of Canada Metal – a fairly routine business problem with the regulatory authorities was treated as a serious threat by Toronto Refiners and Smelters. We believe that the sons sought the best legal advice they could find to protect the reputation of their father.

This meant that the lawyer who represented Toronto Refiners and Smelters – a small company – was one of the most prestigious and outstanding legal counsel available in Toronto. We believe that his instructions were to take any and all possible legal actions to protect the reputation of his clients, and to spare no expense in doing so. Later, we were told that Toronto Refiners and Smelters approached Canada Metal, although it is not clear whether Canada Metal might have also acted on their own initiative in hiring this same lawyer. The companies shared legal counsel in the various hearings and court cases that ensued.

In the early 1970s, most of the controversy revolved around the testing for levels of lead, conducted initially by the Ministry of Environment, its Air Management Branch, and later by the Ministry of Health and by the Toronto Board of Health. The Ministry of Environment began its testing for lead contamination in the soil, probably in 1971. In the area of the Canada Metal plant, it found that the lead levels were six times higher than 600 ppm, which was considered acceptable by the Ministry's standards. The Air Management Branch had found an airborne lead level of 8250 ppm, compared with a "normal" level of 200 ppm. These early studies demonstrated that a problem of lead pollution did exist.

In December 1972, the first Air Management Branch tests were filed with the Ministry of Environment, but the test results did not become public until September 1973, when a community worker managed to obtain the file on Canada Metal. Also in that file was another report by the Air Management Branch from May 1973 that found high levels of lead in the dustfall around the Canada Metal plant. This second report stated that the standards for lead in airborne dust were being regularly exceeded by Canada Metal. A third report, completed in June 1973 and also from the Air Management Branch, measured dustfall and lead content of soils at a school located near Canada Metal. It found that the level of lead content was ten times higher than that from non-industrial areas of the city. The report indicated that the total dustfall from Canada Metal had constantly exceeded the provincial monthly and annual criteria.

In January 1973, however, before all these tests were conducted and finally made public by the provincial authorities, the frustrated citizens formally retained their lawyer from the environmental law group to present their case to the Toronto Board of Health. The Board was responsive to their complaints, for the Board of Health meetings had become the forum for the debate about lead pollution. Because the Ministry of Environment also had jurisdiction, however, questions could – and were – raised by the companies about whether the city had any legal right to issue their abatement orders with respect to matters already being dealt with by the province.

This jurisdictional conflict also undermined the Board of Health's capacity to enforce its abatement orders, or to seek their enforcement through the courts. In fact, the Board of Health issued two abatement orders, one in March 1973 and one in May 1974, and a notice was sent by the Board of Health to both Toronto Refiners and Smelters and to Canada Metal in November 1974 requiring their removal of contaminated soil. In none of these cases could the Board of Health act decisively to secure compliance, nor did it finally refer these matters to the courts.

Regardless of the scope of its legal authority, the Board could – and did – act as an advocate for the citizens in relation to the other city councils and committees and in relation to the provincial Ministries. The Board of Health meetings were somewhat like a court without court decorum, as the lawyers were often present. A city councilor describes one of them, as follows:

> We had a hearing in '72, no '73. It started at ten in the morning and went on till four am without a break. The chairman was a lawyer. The Board sits in a semi-judicial capacity on hearings and you literally can't leave the room. ...The smelters, they didn't like us one bit. The mayor was then hauled into it and he tended to believe their expert. It was our (consultant) versus their expert.

Community meetings were sponsored by the city. At one meeting, the doctor who had been working in close co-ordination with the Board of Health (and who later was a member of the Board of Health) spoke, only to be challenged by a citizen who was apparently briefed by the companies. Ministry of Environment officials were also invited to speak at these meetings. One provincial official described his experience to us, as follows:

> When I first went to the meeting, I didn't get the impression that there were people with their lances out and they were tipped with venom...I saw my role there was that if I could enforce what I thought was a reasonable law, I'd be delighted to do it, because I knew the thing was off the rails. (That) it was a meeting in the area of Canada Metal tipped me off...And it wasn't until I got there, (I found out) there were people in the audience being psyched up to have a go. I didn't know what was coming. Well they got me that night, and it turned out to be Canada Metal.

Toronto Refiners and Smelters now claimed that one hundred thousand dollars had already been spent on its cleanup efforts, and that this amount was sufficient. The company officials also claimed that 99% of the emissions had been contained. When the provincial stop order was lifted in July 1973, the company felt justified in resuming full operations. But the Toronto Board of Health argued that covering the piles of scrap metal with plastic was not a sufficient response to its abatement order. It hired a toxicology expert from the United States to support its case.

Not surprisingly, the citizen group was frustrated by the delays, and the creation of several different testing programs, all with no visible impact upon the pollution. It was angered by the numerous government committees and Ministries dealing with lead pollution, and the fact that the Ministry of Environment appeared to have withheld test results. All the early actions – the measures taken by Toronto Refiners and Smelters, the decisions by the Board of Health, City Council and city building committee, as well as the testing programs of the two Ministries – did not seem sufficient to deal with its view of the severity of the problem. Its perception of delay was heightened when, in June 1973, three youths from the community were taken to hospital with high blood lead levels.

It is worth reviewing the controversy as it had developed until mid-1973. In this first phase of the controversy, two companies, two provincial government departments, the city council, the city's building committee and the city's Board of Health all had become involved. Lawyers had been consulted and retained, for matters that normally would have been handled by citizen delegations to the city, by negotiations with the ministries, or as normal business problems. Some testing had been done; there was ample evidence of a pollution problem. Some individuals appeared to be harmed, or at least to have levels of lead in their blood that were well above those considered normal.

Nonetheless, we have the impression that the problems that caused the Toronto lead controversy still could have been resolved without too much difficulty. The problem with lead pollution was a localized one, involving two companies only. Pollution control was technically feasible, and not prohibitively expensive. If the various government departments and committees had co-operated, and not been engaged in jurisdictional battles, sufficient measures might have been taken to control the lead pollution from the plants.

Several signs existed that the controversy was about to become much more intense, however. The involvement of lawyers, now on both sides of the issue, was an important one, for with legal participants, all conflicts were viewed in terms of the legal actions that might be taken. The lawyer for the companies, we assume, was watching for any possibility that the companies would be charged with negligence and made liable for any damage resulting from their actions. The companies claimed that they were worried about being shut down, and about "being put out of business". A spokesman for the industry told us, "… they were in danger of being shut down." Certainly the companies' lawyer was given instructions to pursue the matter vigorously, for a company official told us:

> (Our counsel) is the best guy in the business...Let's face it, the fact that we have the counsel working with us, I use as a hammer with all the Ministries now. And when we discuss things, they know in the background that if we cannot reach an agreement, if they don't want to listen to logic, if they don't want to approach it in a fair and equitable manner, then you unleash the dogs and say to the counsel, 'here it is.'

All of the legislation was being scrutinized by the company and the environmental lawyers, and was seen as the basis for action or for a sustained critique. We were also told that at least one of the citizen activists saw the situation as follows:

> (Our counsel) viewed the litigation as a way to break the logjam. It was the theory we operated on, that we needed the court case to break it up...Once the ball got rolling, it was the job of the citizen organizations to make sure the ball kept rolling.

The use of experts, as consultants, was another sign that the controversy was about to erupt. The use of experts suggested that the problem to be solved was much more difficult, and more scientifically complex than the initial testing had indicated. It suggested that the scientific status of lead, its toxicity and its effects were all contested matters within the scientific community. It was a sign that the local testing, conducted by the various governmental departments, was seen to be insufficient as a basis for regulatory action.

Finally, the failure of the Ministry of Environment to release its reports, and their subsequent discovery by the lawyer representing the citizen group, cast the Ministry's actions in a very negative light. The involvement of one level of government as an advocate, and another as a regulator also raised questions about the role of regulators. The reluctance of the Ministry of Health to become involved in what was seen to be a serious health problem exacerbated an initial skepticism about the governments' intention to act. By the end of the first phase of the controversy, the situation was rapidly developing to the point where at least the companies and provincial government departments were not trusted.

## The Toronto Lead Controversy (2) – the Case Goes to Court

By mid-1973, three groups had conducted or were conducting research. The Toronto Board of Health did blood testing through a voluntary program aimed at citizens living near the smelting plants. The Ministry of Environment did air and soil sampling, although it became apparent that its testing for airborne lead was incapable of identifying accurately a pollution problem. Lead is too heavy to remain airborne for any length of time.

A third group also began testing for evidence of abnormally high lead levels. This third group was made up of three scientists from the university, working with an institute of environmental studies. The scientists were – and are – highly acclaimed in their academic specialties, none of which specifically include the toxicology of human exposures to lead. As part of their environmental studies institute and as individuals, each had been involved with regulatory and expert committees previously. One member of their institute was a founding member of an environmental advocacy group, and another was a friend of the doctor who was, by now, an activist on the issue of lead pollution, but the scientists who conducted the studies had been consultants, not activists with respect to environmental issues prior to the Toronto lead controversy.

The university scientists began their studies on their own initiative in July 1973, partly in response to the extensive coverage that the lead controversy received through the media. The citizen group heard about this initial research, again through the media coverage, and it asked the group to undertake an independent research study to determine lead levels around the Toronto Refiners and Smelters plant. Their second study was funded by a grant from the city. The two reports by the university scientists were completed in September 1973, but before the second report was published, it was leaked to a community-oriented daily newspaper. The person responsible for the leak was not one of the university scientists, but was instead the doctor from the children's hospital. Once the study was leaked in this fashion, it lost its appearance of independence, quite apart from the methods that were used by the university scientists to arrive at their conclusions.

The second report of the university scientists was of a six month study of the lead in the soil and dustfall around the two smelting plants. It found eighty times more lead in the soil surrounding the Canada Metal plant than in an area selected as a control. It found high levels of lead in the dustfall, and indications that vegetation had been exposed to lead. The university scientists believed that the problem with lead stemmed from fugitive emissions, in other words from the dust, debris and unintentional "leakage" from equipment used at the plants. The lead pollution, while localized, "was so severe as to indicate a potential threat to health". The scientists recommended a prolonged and extensive lead survey program.

The university scientists' reports were presented to a meeting of the Toronto Board of Health in November 1973. Their reception was mixed. The lawyers for the companies, who were present at the Board of Health meeting, disputed the findings. The Ministry of Health consultant concurred with the companies' general assessment of the university scientists' studies, and stated that the problem was less serious at the Canada Metal site than at Toronto Refiners and Smelters. The Air Management Branch officials indicated that Canada Metal had already been asked to install a new dust filtration device. The Ministry of Environment officials did not consider the studies convincing. We were told that the Ministry of Environment's attitude towards the university scientists was "condescending". In contrast, the Toronto Board of Health considered the studies to be very valuable.

The response of the Board of Health, and the growing controversy sparked a chain of other events. They were, as follows: The Ministry of Environment issued a third control order to Canada Metal, requiring compliance by the end of September 1973.[15] This control order mentioned fugitive emissions specifically. Canada Metal was asked to pave its truck roads, install washing devices for its trucks, and wash each truck before it left the premises. Canada Metal was also asked to install roofed enclosures for the lead scrap piles. Fifteen days after the deadline set in the control order, the Ministry officials inspected the plant, and found that the trucks were not being washed. Canada Metal was charged in mid-October – the first formal legal charge by the Ministry of Environment – with a summary offence for violating the control order. The company was fined, and it complied with the control order by mid-November.

A special meeting of the Board of Health was held. A lead toxicology expert was engaged by the city.[16] The Board of Health also asked the city's committee of adjustment to delay a decision on Canada Metal's request for a permit to construct an eighty-five foot smokestack. Both Canada Metal and Toronto Refiners and Smelters had permit applications for smokestacks pending before city committees, and both companies were experiencing trouble in gaining approval.

The companies responded, this time acting together. The companies brought in two scientific experts to speak to the Board of Health meetings. The companies' experts testified in October 1973 that the plants were not the source of the problem with lead, and that the source was probably lead poisoning from paint that contained lead and that been used on household objects in the older homes in the neighbourhoods adjacent to the plants.

The city's Department of Public Health conducted further tests. The Department of Public Health sampled seven hundred people, chosen on a volunteer basis from the neighbourhood adjacent to the Canada Metal plant. The results of these tests indicated that three people had blood lead levels that seemed "dangerously high". As a result, the Medical Officer of Health notified the Air Management Branch of the Ministry of Environment.

In late October, 1973, and partly in response to the Medical Officer of Health's notification and partly in response to this chain of actions and events, the Ministry of Environment issued a *stop order* to Canada Metal, acting under a hithertofore unused section of its legislation. The Minister accompanied the issuance of the stop order with a public statement about why it had been issued, indicating that the Ministry officials had been unable to "isolate the specific sources of the lead emissions", and therefore that they had to shut down the operation.

The day after the stop order was issued to Canada Metal, it was appealed by the company to the Supreme Court of Ontario. The hearing opened two days after it was filed. The lawyer originally hired by Toronto Refiners and Smelters represented Canada Metal. The scientist from the United States who testified earlier that the problems with lead stemmed from paint, testified for the company, and another expert consultant was engaged. The Ministry of Environment – which had issued the stop order that was being challenged – called no witnesses. The stop order was quashed four days after it was issued.[17]

The companies charged that the Ministry of Environment had not included a rationale in the stop order for the plant closure. Indeed it had not, for the closure was posited on the fact that the source of the lead emissions could not be determined. From the perspective of the Ministry, the stop order was a preventative measure. If the cause of the pollution could not be determined yet, it argued, measures should be taken to prevent further damage until it could. The stop order was also a political and symbolic response to growing pressure, a sign from the Ministry that it would deal forcefully with pollutors. As a Ministry official told us, "The second draft was still in the typewriter when the press release had already gone out."

The judge took the position that no emergency or "immediate" danger had been proven, and that the stop order was an arbitrary exercise of authority. The Ministry had not proved "reasonable and probable grounds for such a danger", he stated. Since the legislation indicated that issuance of a stop order was contingent upon there being an immediate danger, the legislation provided no support either for the Ministry's actions or for the position being advocated by the environmental groups.

The Ministry of Environment sustained an extensive attack from citizen groups and from politicians as a result of its failure to introduce expert witnesses. It was said that the Ministry called no witnesses in an effort to lose the case and "get off the hook". A letter to a city council member from an advocacy group is worth quoting at some length, as it illustrates the basis of the position taken by some of the advocates:

> I would point out to members of the Board (of Health) that the socially responsible thing to do, given the weight of evidence over the evening, was to take an (legal) action and to pass the issue on to the courts if necessary. Quite apart from the fact that I think you had sufficient evidence before you tonight to prove your case, one does not have to win every case in order to prove one's point...
>
> When the good doctor reminded the Board that the province had its stop order quashed by Mr Justice Keith, it was not because of the weakness of the arguments that were available to the Air Management Branch (AMB) to support the stop order. It was because the AMB deliberately set out to lose the case or they were totally incompetent. The fact is that they did not present evidence to support their position.
>
> Because of the dangers to the health of its citizens, and the gravity of the environmental crisis, some jurisdictions are now passing legislation shifting the responsibility for the burden of proof to the pollutor...[18]

As might be expected, the Ministry of Environment responded to these and other charges, first by stating that it would conduct more tests in the area surrounding the Canada Metal plant, and that it would speed up Canada Metal's pollution program, giving them a deadline of February the next year. Later the Ministry issued quite a different kind of response. Because the city had refused to provide a permit for the smokestack, the Ministry officials were able to blame the city for the continuing problem, indicating that Canada Metal was in compliance with their provincial requirements. From this point in the controversy, not just the companies but also the Ministry, and indeed all of the government departments were on the defensive.

The stop order and its subsequent appeal received a great deal of publicity. Lead became a political chemical in the literal meaning of the word political. One political party took up the cause, and called for a public hearing; another political party criticized the Air Management Branch and asked the provincial Premier to intervene. Both of these political parties charged that the relationship between the Air Management Branch and the companies had become too cordial, and both questioned the integrity of the regulatory process and of the Ministry of Environment as regulator.

## The Toronto Lead Controversy (3) – Words Become Dangerous

It is difficult to describe the events and actions in the controversy that followed. Little coherence or order is evident in actions, accusations, proposals, and events in late 1973 and during 1974. It is important to provide a picture of the controversy at its height during this period, however, at least as an introduction to the companies' next response. Thus, we have selected a few of these events, and will list them. Our list is not strictly chronological, and we emphasize that each action had many causes.

A child from the neighbourhood of the Canada Metal plant was hospitalized with lead poisoning, and the Toronto School Board, hitherto uninvolved in the controversy, called for an inquiry into the handling of the citizen complaints by the Air Management Branch of the Ministry of Environment. At the same time, however, evidence mounted that the health problem might not be as serious as some believed. Of three other people who had been hospitalized for high blood lead levels after the Department of Public Health's initial testing, it was found that one was hospitalized because of a mistake in the test results, and that another was an employee of a smelting company other than either Canada Metal or Toronto Refiners and Smelters. The picture about the dangers from lead exposure was becoming complicated.

In November 1973, the Minister of Environment chose to back down from his defence of the stop order. He said that the Ministry had acted too hastily, given the now-known results of the Department of Public Health's initial tests. He suggested that he would not now have issued the stop order, and, indeed, that the highway, not the smelters, was the source of the high blood lead levels.

A spokesman from Canada Metal suggested that the lead controversy was simply the result of the "anti-business groups", "mothers' groups" and politicians trying to create an hysterical climate. The President of Canada Metal indicated that it was proceeding with its planned pollution controls, stating, "We can no longer delay these changes, let them sue us or take us to court if they want to".[19] In the legislature, the Minister of Environment – acting in response to these criticisms and charges – said he would seek legislation to force the city to issue the building permits for the smelting company smokestacks. He suggested that the city aldermen were trying to make him "the fall guy". The Mayor of the City then replied that the smokestacks were a "red herring", since the stacks were designed to control emissions of sulfur dioxide, not lead. The Mayor said that the delay in issuing permits was to prevent Canada Metal from having a sense of permanency.

An official from Canada Metal also questioned the scientific validity of the testing methods – which collected the blood from a finger prick. Canada Metal wanted tests done with larger samples of blood that would be taken from each person. Canada Metal was not prepared to be conciliatory. It refused to give its employees time off for the blood tests being conducted by the city's Department of Health, insisting that the blood tests be conducted by the company's own physician and analysed in an independent laboratory. Canada Metal promised to fight any effort to close or move the plant, and brought a third expert, this time from England, to visit officials and discuss the problem of lead pollution.

New tests were then made public by the city's Department of Public Health, indicating that a number of citizens had elevated levels of lead in their blood. Further tests by the university scientists were also released. These latter tests indicated that the lead pollution around the Canada Metal plant had "drop(ped) dramatically" during the five days when the plant was shut down. This was the first evidence that could be used to link the emissions to the hazard – linking the cause to the effect. The university scientists stressed – somewhat defensively – that they were not medical doctors, and that they could not state whether or not the lead pollution around the Canada Metal plant was hazardous to health. The Ministry of Environment claimed that these reports by the university scientists were calculated to raise alarm.

Finally, at a joint meeting of the Board of Health and the city's development committee, Canada Metal and Toronto Refiners and Smelters were given permission to proceed with their pollution abatement programs, but not with the construction of their smokestacks. The Board of Health – whose legal jurisdiction with respect to the enforcement of standards was in question – also set a deadline of February 1974 for the cessation of the pollution, indicating that it would seek a court order if the pollution problem persisted. In December 1973, the Toronto Board of Health issued tougher guidelines – standards for blood lead levels – concerning pollutants.

Events took quite a different turn when the companies decided to take the offensive. First, Canada Metal challenged the appointment to the Board of Health of the doctor who been its advisor since the beginning of the controversy and of the alderman who had advocated the citizen's cause previously. Canada Metal suggested that their activism precluded their responsible representation on the Board of Health. Second, Canada Metal cut back on its operations, and claimed that cleaning up the debris in its yard had been responsible for the high level of lead inside the smelter and the disturbing results of its worker testing program.[20] Third, the two smelter companies went to court in January 1974 to prevent the broadcast of a radio program about the Toronto Lead controversy. This was followed by a series of court actions based on charges that the program was biased.

The program was called "Dying of Lead". It was broadcast in the eastern regions of Canada. Before it could be broadcast to central and western Canada, the companies sought and obtained an injunction. In response to the injunction, the journalists deleted the two offending segments of the program (each less than a minute in length), left silence in their place and explained the situation in an introduction added to the program. The message was not lost on the audience, or the politicians. The Minister of Health immediately ordered mandatory monthly blood tests to check on the employees at Canada Metal, acknowledging the role that public pressure played in his decision. A parliamentary leader also used the occasion to repeat his earlier request for a public hearing.

The message was also not lost on the smelting companies which immediately returned to court to claim that the program producers, the government-owned broadcasting company, a local newspaper, a citizen group and several of those interviewed should be charged with contempt. The company claimed that each had violated the injunction by issuing information via news stories and leaflets. The companies asked the court for a permanent ban on the program which was said to reflect "scandalous irresponsibility". The companies cited a new study, a review conducted by one of the national government-funded research councils, which had suggested that lead did not pose a serious health hazard in the Canadian environment. At the same time, Canada Metal also distributed a pamphlet to residents. The pamphlet dismissed the threat from lead pollution.

Later the company charged the broadcasting company, five of its employees and a local newspaper with bias and sought recompense of fourteen million dollars. It also sought an injunction against the Board of Health, two of the city aldermen, and the doctor from the childrens hospital to prevent any of these groups and individuals from dealing with the lead issue any further. A libel suit was also brought against several media. Later, a suit for "spreading false news" was also threatened.

We cannot stress too strongly how unusual such actions were in the Canadian context. We know of no other example where charges of bias were laid in response to an important public issue against individuals and against an agency of government. There have been instances of libel actions against media, but no instances (to the best of our knowledge) in which those connected with such a media program have been successfully forbidden from speaking generally about the public issues that the program raised.[21] We know of only one instance in Canada in which a complainant sought to use the courts to prevent a media program from being rebroadcast, and in that case, the complainant was appealing a decision of the broadcasting regulatory agency and was unsuccessful.[22]

In February 1974, the Supreme Court lifted the ban on the radio program, but the companies sought a new injunction just as the program was about to be aired for a second time. The first of the charges came to court in March 1974. One person was cleared of the contempt charges but was ordered to pay eighty per cent of the court costs. Eventually, the judge found in favour of the companies. A newspaper reporter and two broadcasting journalists were fined for violating the injunction. These fines were later suspended, on the condition that the journalists pay half of the court costs.

The company then pressed its legal charges against three Board of Health members. Although members of a government agency, these individuals did not have legislative immunity. The company also sought subpoenas for the twenty reporters who had provided press coverage of the Board of Health. The court denied the subpoenas, but the Board of Health members were found guilty, and the city government eventually had to pay their court costs. The legal costs of all the defendants were in the magnitude of $200,000. The final judgement on matters stemming from the radio program was issued in 1978. The broadcasting company was ordered to pay one of the company's experts $20,000 for defamation.

There is no question that the radio program was worthy of the companies' concern. The program suggested that the smelters had "bought" favourable medical evidence, that they had concealed evidence from those experts and that they had misstated the amount of money spent on pollution control. Of particular concern to the companies was a statement within the program by a company expert which indicated that the reports by the Ministry of Environment had not been reviewed before testimony was given in the stop order hearing. Nonetheless, the injunctions and accusations of bias and false news had a chilling effect on the participants in the controversy, particularly because these legal actions were so unusual. Every individual, not just the companies and the government departments, now had a good reason to be on the defensive.

### The Toronto Lead Controversy (4) – Studying the Problem

Initially, little doubt had been expressed either about the validity of the testing for levels of lead or about the significance of the results. At the same time, it would be a mistake to think that the initial research was either well planned or properly co-ordinated. As a Ministry of Health consultant said later:

> Things had been added along the way, bringing in paint, looking at soils, and then asking questions about kettles and drinking water and all kinds of things. There wasn't a master plan and nobody ever thought of what they should do with the data...Well, they (the Board of Health) turned around at some point and said the city can't possibly analyse it...And so the Province was brought in to supply the where-with-all, so we could analyse the data. Well as you can imagine, it was a dog's breakfast.

By mid-February 1974, a new theme was evident in the press coverage. A story in the *Toronto Star* was headlined "Even Experts Cannot Agree on Lead Poisoning", and it discussed the problem of scientific uncertainty at some length.

This next phase of the controversy was characterized by a flurry of new research and by the release of the reports from a number of new expert committees. The Board of Health, for example, brought in a new American expert, who was asked to examine the existing test results from the university scientists' studies. He consulted with the various expert committees and produced his first report by March 1974. He proposed a study to provide evidence that would link the activities of what was now three smelting companies and the elevated blood levels.

In this proposed study, attention was to be paid to the effects on all the neurological and behavioral problems experienced by children who had been exposed to lead, one of the more controversial aspects of lead's toxicity. The city council voted to ask the province for $250,000 for the study. The province refused. The Board of Health then approached the city council for the funding of a more limited study involving testing citizens for their blood lead levels. This second study was funded and carried out, but not by the Chicago research team. The Board of Health also used several other experts as consultants during this period.

In late 1973, the Ministry of Environment commissioned an expert group to examine the information on lead exposures and dangers. This committee was called the Working Group on Lead. It was made up of officials from both the Ministry of Environment and Health, and an as-yet uninvolved scientist from the local university. Then, in March 1974, the Ministry of Health also commissioned an expert committee to review the test results and studies on lead pollution in terms of the possible health effects of the pollution. The Robertson Committee, as the Ministry of Health's expert group came to be known, was made up of three scientists, one of whom was an environmental activist, and who was active in a federal science advisory council.

In turn, the companies hired their own scientist to conduct research. According to a letter from a city councilor to the city's medical officer of health, the Ministry of Health, the university scientists, a scientist from the Ontario Research Foundation and the Toronto Board of Health each had raised questions about the methodology of the proposed study. The company-sponsored study was conducted nonetheless. and the scientist involved responded to all of the various criticisms of his study and report in a series of letters in December 1974. He concluded that neither the smelters nor the expressway were the source of the high lead levels – paint was the source of the lead problem. The companies also commissioned a second expert to do a peer review of the university scientists' research. This expert found problems in the university scientists' research, but his review was challenged in turn by the university scientists.

In June 1974, The Working Group on Lead presented its interim report to the provincial legislature. In one of its submissions, written by the Ministry of Environment, it said that in a sample of forty-five smelters in Ontario, more than half were causing lead pollution. The three Toronto plants were among the worst pollutors in the province. The release of the interim report fuelled criticism of the various provincial Ministries. Indeed one legislator suggested that the Ministry of Environment "should be scrapped" because it was "rotten". The Ministry of Health officials were accused of deliberately sitting upon the report, to which a Ministry official apparently replied that the Working Group on Lead's interim report "got lost in the shuffle" of papers on his desk.

The Working Group on Lead presented its final report in August 1974. The Working Group had consulted with both of the consultants that were used by the Toronto Board of Health earlier, and in its report it referred to the university scientists' research, the report by the expert who designed the proposed study for the Board of Health, and the reports of the Air Resources Branch (formerly the Air Management Branch). The Working Group report was critical of the smelters, and concluded that the smelters had been major contributors to the lead pollution. The report cited problems with the dustfall and stressed that the fugitive emissions were the main source of lead pollution. It called for more stringent standards, and the removal of the topsoil from near the smelters.

The Minister of Environment accepted the report from The Working Group on Lead, but then said that he would wait for the results of the expert committee commissioned by the Ministry of Health before acting on its recommendations. In September 1974, however, Canada Metal was ordered to stand trial on charges that it had failed to meet the conditions of its control order by the required deadline. And in October 1974, a third company (four companies were eventually involved) went to trial, was found guilty and was ordered to pay a two thousand dollar fine.

In March 1974, the university scientists presented a report on proposed standards for air quality to the city's Department of Public Health. The report was reviewed by a Ministry of Environment official in April, who stated that it was less adequate than the Ministry's own assessment, completed in December 1973, and that it "contain(ed) nothing new".

Nonetheless, the university scientists' report was the occasion for a sharp exchange between the companies' lawyer and the Board of Health. The lawyer's letter to the Board of Health stated that the US Environmental Protection Agency could not be relied upon for a lead standard because EPA had no standard for lead and because EPA had not established the link between a standard for airborne lead and human health. The problem with this exchange is that the university scientists' report contained no reference to EPA, but the original Ministry report did. In the thick of the controversy, it was difficult to distinguish who was responsible for which assessments and recommendations.

At the end of October 1974, the final report by the university scientists before the commencement of the environmental assessment hearings was released.[23] This report concluded that area adjacent to Toronto Refiners and Smelters had higher lead levels than that around Canada Metal, although the lead content in the soil and dustfall in both cases was above normal. The report also suggested that the lead levels decreased in proportion to the distance to each plant, again providing some substantiation for the claim that the smelters were causing the lead pollution.

Finally, the Robertson committee issued its report. This report provided a description of the jurisdictional problems between the various Ministries. It suggested that the companies and governments had been unwilling to act. Its main conclusion was that residential areas and lead smelting facilities should not be located in close proximity to each other, and indeed that the smelters should be moved immediately regardless of the political or economic ramifications of doing so. The Robertson committee also endorsed the Working Group on Lead's recommendations, but its endorsement was a weak one, and the Robertson committee report offered no strong recommendations of its own on the question of standards.

The impact of these numerous studies and committee reports should not be underestimated. What was a fairly simple problem, the fugitive emissions of lead in particular, was becoming a very complex one. What was generally agreed upon, the toxicity of lead, was now a subject of scientific controversy. Each new study called for additional research, and served to muddy the waters further. It is not surprising that the recommendations of each expert group differed, but it is worthy of note that the study teams could not agree on an appropriate methodology for a seemingly simple piece of research. The assessment process to this point in the Toronto lead controversy provided little assurance that science would contribute to the resolution of the debate. Little wonder then that the environmental assessment board hearing, which followed, did not do much better.

## The Toronto Lead Controversy (5) – The Hearing Acts as a Court

The next stage of the controversy began in November 1974 with the announcement of public hearings under the auspices of the provincial environmental assessment board. The environmental assessment board was empowered to consider the various reports and research of the other expert committees, including the two provincial expert committees that had already reported. Any action on the problem of lead pollution by the provincial Ministries, other than those already in motion, was to be postponed until the hearing board issued its report. The creation of the environmental assessment board did not stop the Toronto Board of Health from ordering the smelters to remove the soil from around the plants, or the smelters from refusing, however. This new conflict was also sent to the environmental assessment board to resolve.

The hearings of the environmental board opened in January 1975. The first question to be dealt with was one of procedure. The companies sought the right to cross-examine all witnesses, an unusual procedure in the Canadian context. The Ministry of Environment and the environmental advocate groups objected, but the companies won, and the hearings assumed the atmosphere of a court as a result. By the time the board was ready to hear from the advocate groups, the groups were quite frustrated both with the scheduling of the inquiry and with its approach to assessment. Some of the advocate groups later withdrew from the environmental assessment board hearings.

The Board of Health was initially represented by the legal counsel. He worked with the same environmental advocate group that had represented the citizens at the Board of Health two years earlier. The Board of Health's expert consultant was one of the university scientists. On the fifteenth day of the hearings, the Toronto Board of Health withdrew, because it was unable to get sufficient financial support for its legal counsel. The Board of Health later returned to the hearings, and its final presentation was made by its Chairperson.

Both the industry trade association and university scientists were cross-examined in some detail. In addition, a succession of expert witnesses from various sides of the conflict dealt mainly with the health aspects of the lead pollution. Final arguments were submitted by the counsel for the Ministry of Environment, and for the companies, a member of the Toronto Board of Health, two of the university scientists and the president of a fourth smelting company. The hearings lasted forty-six days, scheduled over a nine month period.

The issues discussed were, as follows: the difference between regulations and guidelines, the sources of the lead contamination, the long-term effect of low dosages of lead, the immediate effects of lead on children, the right of the public to access to the relevant information produced by various governmental departments and agencies, the degree to which the public could and should have participated in regulatory proceedings, the problem of proving human harm and the "burden of proof" in science and in legal proceedings, and finally the problems in jurisdiction.

The cross-examination by the companies of the other witnesses was, in the words of several observers, "brutal" and "aggressive". The hearings themselves have been described by a government participant as a "free for all". This observer commented on the situation, as follows:

> It was a circus. (The Ministry) did not do a bad job before the court, but the hearing was very badly conducted and was much longer than it needed to be. And the appearance of fairness was not maintained at the hearing because of all the public... The only thing that was visible to them was the personality of the (counsel from the companies) dominating the proceedings. The Board cowered before him. The Ministry was lashing at him. But nobody was really powerful enough to bring the process under control.

Midpoint in the hearings, yet another task force was set up. This fourth expert committee, The Lead Data Analysis Task Force, worked with the Ministries of Environment and Health and members The Working Group on Lead. It was mandated to co-ordinate and analyse existing data from the various studies and other expert committees. The Lead Data Analysis Task Force conducted the epidemiological analysis of the existing data with the Toronto Department of Public Health. It invited designated representatives of all the participant groups including the Ministries, the companies and the Toronto Board of Health, to sit with the Task Force as observers.

The Lead Data Analysis Task Force concluded that there was a strong relationship between the results from the blood tests and the results from the testing of soil lead levels. It concluded that the people living near the smelters had a higher risk of an elevated blood lead. The Lead Data Analysis Task Force recommended the removing of the soil around the plants. Although The Lead Data Analysis Task Force report was prepared ostensibly for the environmental assessment board, and available by September 1975, apparently it was not examined by the board before the report was produced, and its data were not included in the board's report. The reason was said to be "a bureaucratic foulup". The report of the environmental assessment board indicated that reliance should be placed on the Lead Data Analysis Task Force as soon as the report became available.

The report from the environmental assessment board was released in April 1976. At the same time, the report from The Lead Data Analysis Task Force was made public. The environmental assessment board report dealt with the recommendations of the Working Group on Lead and the Robertson Committee. In general, however, it concluded that there was no serious health problem with lead pollution stemming from the smelting operations in the Toronto area, although soil and dustfall levels were higher near the plants. The report recommended more stringent standards for blood lead levels and routine blood tests for children and for pregnant women living in the smelter areas. It recommended clearer definitions of standards, criteria and guidelines, and recognized the problem caused by the lack of standards for fugitive emissions. Finally, it recommended increased stringency of the lead standard for occupational exposure, but did not stipulate the appropriate standard to be applied.

The Toronto lead controversy did not end with the report of the environmental assessment hearing, but something had changed. At first, the problems with pollution continued, and the various governmental departments and boards continued to question each other's intentions. A control order was issued against Toronto Refiners and Smelters in April 1975. The university scientists felt that this control order constituted a significant step in controlling the sources of pollution from the plant.[24] Four charges were also laid against Canada Metal for pollution offenses.[25] The company complained, in response, that it had already spent almost $300,000 on pollution control equipment, but pleaded guilty on two of the four charges. In finding against the company, the judge called the company a "good corporate citizen" and suspended the sentences on the basis that the damages were minimal and that there was no possibility of reoccurrence. The Ministry of Environment appealed, and this time the Ministry won its case.

The university scientists completed their last report on the lead pollution in 1980, and each has now taken up other scientific interests. The Toronto Board of Health and other city agencies, the Air Resources Branch and the Central Regional Offices of the Ministry of Environment and the Federal departments are all still involved, as are some of the citizen groups, and a "Lead Liaison Committee" was established in 1979 – it is called the Environmental Liaison Committee now – with participation from the citizens, the lead companies, the local member of parliament and the Ministry of Environment. A major conflict did occur – and continued for some time – about who should pay the cost of removing contaminated soil from near the smelting companies, and how much soil should be considered to be contaminated. The issue for most of these groups is, now, standards for lead in gasoline or the noise and dust problems related to, but not exclusively created by the smelting companies.

Labour standards were introduced after new legislation was passed in 1978. A lead monitoring system was also established under the provisions of the provincial *Occupational Health and Safety Act*. These last changes were also the result of the recommendations made by yet another inquiry, a Royal Commission on the Health and Safety of Miners. The *Environmental Protection Act* was revised in 1980, to tighten up its mandate and increase the provisions for the enforcement of its standards. The *Public Health Act* was revised with the introduction of the *Health Protection Act* in 1983.

## Discussion

### *Political Chemicals:*

It is useful to the understanding of the Toronto lead controversy to introduce a new term, "political chemical". Political chemicals are chemicals that act as a lodestone for a broad debate about chemical regulation. Political chemicals are controversial, but they are not always the subject of a widespread public debate. For example, pentachlorophenol, another of our case studies, is a political chemical because of its association with Agent Orange and dioxins. Agent

Orange and dioxins are the subject of an extensive public debate; pentachlorophenol is not.

The importance of the term "political chemical" is that it responds to some questions arising from the Toronto lead controversy. For example, if the toxicity of lead was not really in dispute, and almost everyone agreed that pollution control measures were necessary, why did lead and the lead standard become so controversial? The answer is that the Toronto lead controversy also focussed on jurisdictional problems in regulating chemicals, the underdeveloped state of environmental regulation at the time, and misunderstandings about the differences between legal and scientific investigations. The Toronto lead controversy provided the terrain, a "battleground" on which these problems could be addressed.

We would argue that once it becomes "political", a chemical can no longer be assessed dispassionately, either in a court of law or by scientists. Everyone and every report is viewed in terms of how it "lines up" on various sides of the conflict. The university scientists in the Toronto lead controversy were not originally advocates, but they became advocates when lead became a political chemical. They would not have wanted it to be otherwise, yet their reports could have been produced by any similarly qualified scientists. Because their initial reports were released to the press by a known advocate, and because lead was a political chemical, their studies were seen to be politically motivated, and biased. This was upsetting, for each university scientist believed that he was capable of dispassionate investigation even if he also held strong opinions on the matters being investigated.

The term "political chemical" was first coined to deal with the findings of another case study.[26] This was a study of the pesticide captan. A brief review of the issues in the captan case will be useful to show why it was necessary to introduce the term "political chemical" in the study of mandated science.

Captan is one of a number of chemicals – 134 in all – whose regulatory status was thrown into doubt when it was discovered that the chemical had been approved for use on the basis of fraudulent laboratory tests. Several articles appeared in the media naming captan as a chemical worthy of concern. The status of captan quickly became a matter of debate in the Canadian House of Commons, and the centre of a dispute between two Cabinet Ministers and at least two government departments. For the first time in Canada, a consultative committee was appointed to make recommendations with respect to captan.

One of the research team on the mandated science study was appointed to the consultative committee. The experience provoked some questions. Why captan? Why should this pesticide – among the 134 other chemicals whose regulatory status was thrown into question by the discovery of faulty laboratory tests – have attracted all the attention? There was no evidence that captan is the most dangerous chemical on the list. Scientific conflict existed about many of the other chemicals on the list. Why did the spotlight fall on this particular pesticide, and why did the debate about the method of registering pesticides in Canada take place initially and primarily within the context of a debate about captan?

The answer proposed by us at the time was that the captan case provided an opportunity lacking elsewhere for a public debate that was sought by a number of industry and advocate groups. It also provided an opportunity to resolve an on-going conflict between several government departments. We suggested that the choice of captan was serendipitous. Captan was simply the subject of several American reports, and it was discussed (probably as a result of the American reports) in a couple of Canadian newspaper articles.

We would argue that lead became a political chemical for many similar reasons. One was serendipitous, the desire of the family who owned Toronto Refiners and Smelters to engage the best possible legal counsel regardless of the cost. With the participation of a high profile lawyer, the issues were seen to extend beyond the actions of Toronto Refiners and Smelters. The second was the engagement of more than one citizen group, concerned with similar problems at different sites and involving different companies.

The third was the jurisdictional complexity of government relations – which is more typical than is usually acknowledged. This complexity gave rise to a situation in which one governmental body could – and did – assume the role of advocate in relation to the others. It also ensured that the competition and conflict between the various government departments and levels of government would have its own dynamic, that each department would be defensive of its activities and that each department would act as a prod for the others.

Fourth, we believe that any controversy has its own dynamic and momentum. Once the situation in Toronto had reached the point where the controversy was established, it was very difficult for any of the parties to agree to its resolution. The conflict was further entrenched and broadened, as groups found new reasons for their growing lack of trust in the regulatory process and in the efforts of government departments to act in any politically neutral manner.

Fifth, the fact that each scientific study stimulated others was both a cause and an effect of the status of lead as a political chemical. Each group sought the definitive research that would resolve the issues in its favour. In fact, more research seemed to complicate the debate further, a point we will address at some length in the next chapter. Finally, the complexity of the science did not obscure the fact that lead is toxic. The simplicity of the issue, and the achievement of results – however meagre at first – as a result of advocacy, prodded all groups to further action.

We would argue that the term "political chemical" is a very useful one for the study of mandated science, and for understanding some aspects of standard setting. In the case of standard setting, the emergence of political chemicals is easily perceived as threatening by standard setting organizations. They believe that the controversy accompanying the assessment of a political chemical interferes with the quality of its assessment, since many seeming extraneous variables are introduced. As a result, and because they seek to develop a comprehensive list of standards, they seek deliberations that are public only in the most limited sense of the term. The meetings of ACGIH and of CCPR are open to the public, but few – indeed no one – attend. Both of these organizations would be extremely reluctant to introduce public hearings, fearing that the availability of a public forum would result in the creation of many political chemicals, and slow their work to an intolerable degree. They base their argument against public hearings on the example of regulatory standard setting that has indeed spawned a number of political chemicals and relatively few standards.

For mandated science, the existence of political chemicals complicates the relationship between science and public policy in two ways. First, it is difficult to limit the examination of scientific information to the chemical at hand, once the debate encompasses general issues about chemical carcinogenicity or regulation. Secondly, debates about specific chemicals take on the character of a "set piece" in a familiar stage play, as the same issues are raised about each chemical under consideration. The "set piece" nature of the debate is disguised by the fact that the tribunal or committee is considering only one chemical at a time.

Participants in more than one hearing recognize the themes, the structure of arguments and even the parade of expert witnesses from one chemical to the next. To the extent that each chemical is considered to be – or potentially to be – a political chemical, the staged nature of the debate can be justified by each of its participants. If a forum is lacking for the more general debate about chemical regulation, the opportunities provided by any specific chemical assessment cannot be ignored.

Political chemicals also arise because the scientific information necessary for a decision about standards or regulations is often inconclusive, contradictory or uncertain. There is a limit to which a court, tribunal or committee can pursue any scientific issues, if the research does not support a conclusive interpetation. Participants in these hearings are cognizant of the problems faced by tribunals. Their attempt to broaden the debate is, in part, a strategy of persuasion, to deal with the inconclusiveness of the scientific data. Thus, it is not surprising that industry, government and advocate groups all choose, from time to time, to broaden the debate and to establish a particular chemical as a political chemical.

### *Science in a court room setting*

Because courts are seldom used to conduct or review chemical assessments in Canada, the Toronto lead controversy is unusual. Much of what might be taken for granted in other jurisdictions was contentious in the Toronto case. The judge who dealt with the stop order provides a demonstration of some legal problems that can occur when science enters a courtroom setting.

It is important to establish the groundrules that were operating in this instance. In Ontario law at the time, and because of the type of legal action involved, the burden of proof lay with the Ministry to defend its actions. The Ministry of Environment is also quite frank about its inexperience at the time when it issued the stop order. It saw the stop order as a preventative measure. As a Ministry official stated:

> It was my judgement as to whether we should close the plant. And what we had was a series of blood leads, which indicated to me in my judgement that if we allowed the plants to continue, all that we could expect was a situation that I felt to be less than acceptable.

Preventative and symbolic responses by government departments are usually not adequate defences in a court of law for actions that impinge upon the established rights and interests of other parties. Even if the Ministry had called its own witnesses and defended its stop order vigorously, the Ministry was in a difficult situation. The same Ministry official describes the situation as follows:

> Okay, but look at the timing. Friday we issued a stop order. Monday or Tuesday we were in court. Now you get the court order over the weekend, with a witness brought in from the States. You wouldn't expect that. I doubt whether anyone would have an organization that was capable of responding as quickly as that. Now, we didn't expect that.

These difficulties were compounded by the judge's understanding of the relationship between science and the legal process. The judge said that if the Air Management Board wanted to apply a test to determine whether a plant should be closed, it should be an objective test, not a subjective one.[27] The judge saw the distinction between a subjective and objective test as a legal – not scientific – test of evidence, and as relating to the specific powers granted to the Director under the *Environmental Protection Act.* As he interpreted the *Environmental Protection Act*, it was not possible for actions such as the stop order to be based solely upon the discretionary judgement of the Ministry (as it might have been with other legislation). He stated:

> ... I have no doubt that the Environmental Protection Act, 1971 makes it perfectly clear that a Director, acting under the powers conferred upon him by s. 7 or 12 or both, in issuing a stop order must act judicially and, hence, the test that he must apply in order to reach an opinion based on reasonable and probable grounds, is not a subjective test but an objective one.[28]

The judge was wrestling with the problem of whether scientific observations, such as those he had before him, could provide the kind of information that would be required to support a stop order. He stated:

> I can only say that while the reported blood lead levels in the three cases relied on may well be unsafe, dangerous to the life of the individuals and requiring prompt medical attention, it is not a reasonable ground to assume that their condition was caused by the proximity of the Canada Metal plant to their homes having regard to the fact, as shown in para 6 of the above-quoted affidavit, that 725 persons were tested and 722 were in effect found not to have unsafe blood lead levels...In other words...only one person at most had a high blood level out of the 725 tested, that at the time could not reasonably be accounted for other than by the presence of the Canada Metal plant.[29]

Because performance standards were used, the Ministry had to demonstrate that the emissions from the plant were causing the harm. He had to hear evidence that harm had occurred. Most important, he had to hear evidence that a link existed between the specific emissions from the specific plants and the specific harm to the citizens.

The tests conducted by the Ministry, however worthy, were insufficient, because they did not even seek to establish the link between emissions and harm. As well, the lead in the dustfall and blood could have come from many sources, and indeed could have followed a number of different pathways. Circumstantial logic, and partly controlled experiments (such as those conducted by the university scientists) provided evidence that the problem was worst in the neighbourhoods adjacent to the plants. These neighbourhoods were also located near a major expressway, and they were made up of older homes in which lead pipe plumbing might still be used. The homes in these neighbourhoods were even likely to have contained furniture old enough to have been painted with paint containing lead. With a narrow interpretation of the law and application of legal norms, the judge could easily conclude that the Ministry had not established cause for its actions.

The task facing the companies' experts was both simple and of little scientific significance. From a legal point of view, all that these experts were required to do was to introduce some doubt as to the exact source of the high lead levels. As we have said, this was an easy task, because of the age of the housing and the proximity of the smelters to the highway. As the judge noted when he quashed the stop order:

> I have not referred to the evidence of Dr. Sachs, an expert on lead poisoning called by the applicants after I granted leave to adduce oral evidence. I only comment that Dr. Sachs' evidence pointed up the great danger of jumping to conclusions without proper study.[30]

From a scientific point of view, it is entirely unexceptional to say that the lead could have come from another source than the smelters. It was no more or no less scientific than a counter-claim that the source of the lead was the emissions from the smelters. For equally in both cases, neither observation nor measurement could easily produce more than circumstantial evidence as to the source of the problem, the link between the emissions and the harm being caused.

There is a lesson here for both standard setting and mandated science. For standard setting, it is important to note that *regulatory* decisions created the problem for the judge and for the Ministry of Environment. First, because the Ministry was working with performance standards, it had to establish a link between exposure and harm. In a natural setting, this link is difficult to prove, especially if the standard of proof is a legal one. In this particular instance, the Ministry of Environment would have been in a better position if it used prescriptive standards, and implemented these standards with an active inspectorate and system of record-keeping. We will return to this point later, and discuss the implications of the choice between prescriptive and performance standards.

Second, the failure of the Ministry of Environment to document a record of violations of program approvals and control orders prior to issuing the stop order also contributed to its problems in court. As a Ministry official told us:

> It is easy to assume that there is a system to standards when there is not. If every time you set occupational standards, you had a hearing in front of you, imagine! You'd have to do a lot more thinking and qualifying. We weren't involved (at the time) in standard setting. We were involved in a defence.

The Ministry was working within the constraints of a political process. In this instance, these constraints resulted in inadequate attention being paid to the standards and in an imperative to act, even with insufficient information. The judge did not recognize these political constraints as legitimate, particularly given the nature of the legislation governing the stop order.

For mandated science, the Toronto lead controversy illustrates some dimensions of the relationship between science and the legal process. The legal standards of proof, the legal meaning of "subjective" and "objective", and the existence of legal "rights" are quite different from the norms associated with science. The judge applied a legal test to the evidence before him, as he was required to do. He also applied a layperson's commonsense. His assessment of the science was unscientific.

For example, from a scientific standpoint, the statistics he quotes neither prove, nor disprove the contention of the Ministry of Environment. The judge seemed to believed they could. The source of the lead could not be pinpointed with accuracy, even if scientific evidence supported one view. The existence of contrary scientific views was not, in itself, sufficient to call the university scientists' research into question, however. The judge seemed to believe it was. The rights of the companies with respect to the stop order had no bearing on the scientific merit of the Ministry's case. The judge paid a great deal of attention to the rights of the parties against whom the stop order had been issued. With either a legal or a commonsense perspective, it was easy for him to mix the scientific and legal issues and to evaluate the research using legal and not scientific criteria.

*Science and the Environmental Assessment Board Hearings*

The environmental assessment board hearings were also conducted in a court-like manner. By the time the environmental assessment board hearings were held, everyone was acting with a legally inspired caution, and the use of cross-examination in the hearings compounded it. We would suggest that the main burden for turning the hearings into something resembling a court rests with the company lawyers, however.

The lawyers for the companies pursued two basic themes throughout the months of sustained cross-examination. One was the difficulty of drawing firm conclusions about the sources and effects of lead contamination, in view of the diversity in types and pathways of human exposure to a widely used substance. The other was uncovering what can only be called "regulatory sloppiness" in setting environmental standards, that is the relationship between the scientific understanding of lead pollution and its effects, on one hand, and the bureaucratic response to that understanding in the decision making procedures employed in setting and enforcing standards, on the other.

The companies' lawyer came to the hearings with the perception that scientists are "scoundrels", who like to use "neutrality and objectivity" as a cover for their own commitments to "various ideas and causes". While his own task, as lawyer, was to protect "ideas and causes" – the interests of his clients – he presented the companies' witnesses as just what they were, scientists who had authored publications and who had also done work on behalf of the lead companies; the results and quality of that work was in the public domain, subject to refutation according to the normal canons of scientific review. And in cross-examination of the Ministry's witnesses he attempted, successfully in our view, to undermine any implicit claim to neutrality on the part of publicly funded academic researchers.

In directly accusing one of the university scientists of bias, the lawyer was pointing, not so much to the studies themselves, but to the regulatory conclusions that the university scientist himself had drawn from them, quite frequently, during his period of association with the Board of Health. That is, the university scientist had made proposals about what should be done to reduce the health risks from lead pollution for residents in the areas near the smelters.

The lawyer's point was simple. These regulatory conclusions and proposals were not the university scientist's area of expertise, and yet he was attempting to act as an expert with respect to them. The lawyer for the companies claimed that his intention was not to discredit the academic reputation of the university scientists – a debatable point – but only to put all of the science regarding lead on the same footing – so as to turn the spotlight away from science and onto another target: environmental standards. The lawyer was able to make the accusation of bias credible, because only the university scientists had openly made recommendations with respect to regulations, while the companies' experts had not.

The second theme came to the fore quite early on in the hearings, in the lawyer's cross-examination of a Ministry of Environment official, a meteorologist who was responsible for monitoring air quality indicators. This Ministry official had actually chaired many of the Ministry's meetings where there were extensive discussions of setting and changing air quality standards, but the Ministry official stated at the hearing:

> You see, I am not developing standards, and I am not in a position to discuss whether or not a standard is being enforced. It is not in my expertise.

Asked for an explanation, he continued:

> In the group of people who are producing the standards, we have a sort of a committee of a medical man, Phytotoxicologist, and an engineering person and myself, so I actually chaired some of these meetings. We discussed this but the actual standards were the result of the Phytotoxicology and the medical person, with respect to contaminants that affect either health or vegetation.

The companies' counsel then replied:

> All I want to be assured of is that at some stage during this hearing I will have the opportunity to cross-examine the person who will accept the responsibility for setting this (air quality) criteria and if you are not that person, I don't want to waste the Tribunal's time...Are you that person or aren't you?

Given the actual capacity at the time within the Ministry to set standards, its defensiveness, its reliance upon ACGIH (which some Ministry officials might not have known about), and the lack of scientific conclusiveness generally attached to deliberations about standards, we are not surprised that the answer of the Ministry official was negative and that the lawyer never did locate that person.

In a court or court-like setting, issues of liability lie just below the surface of the discussion. Any responsible lawyer representing the companies – as the companies' lawyer undoubtedly was – would have directed the questioning so as to ensure that no hint of a charge relating to negligence of the company or its liability was left unchallenged. To the extent that such a lawyer controlled the flow of the debate – as he did in this instance – the hearing itself took on the orientation of a court.

In a court, as we have noted, one would have to prove that harm had actually occurred, and that the harm was a direct result of exposures from a specific source of contamination. The burden of proof would be with those who sought to prove that damage had occurred. This was not the mandate of the environmental assessment board when it was first commissioned. It is not generally the mandate of an expert panel or an inquiry. Nor is this perspective reflected in the board's final report. This is, however, exactly what happened in the environmental assessment board hearings. For all intents and purposes, the hearings themselves became a court of law. The Ministry that commissioned the environmental assessment board was quite unprepared for the result.

In mandated science generally, and in standard setting in particular, it is common to make a distinction between expert committees (task forces, panels, inquiries and commissions) and the courts. Expert committees are designed partly to avoid the limitations of the legal process, and to facilitate scientific debate about public issues. The experience of the environmental assessment board suggests that the distinction between expert committees and the courts is not as clear as is believed. If those engaged in debate before an expert committee conduct themselves as they would in a court of law, and if issues of negligence and liability are raised, the meetings of expert committees turn into something resembling a court.

As the Ministry of Environment found out, a public hearing that both does and does not resemble a court combines some of the worst aspects of both. For example, the testimony of the environmental assessment board illustrates how far an inquiry or expert committee can stray from a scientific discussion of issues. At the same time, the environmental assessment board hearings were not a model of due process. This is a critical observation in the study of mandated science, and we shall return to it in the last chapter of this book.

Chapter Six

# An Economic Poison
# Case Study Four: Pentachlorophenol

*with Edwin Levy*

At least until recently, the product sold by the local hardware to protect wooden decks against deterioration probably contained pentachlorophenol. A warning on the cans says that pentachlorophenol is dangerous. No one seriously questions whether pentachlorophenol is dangerous. Like many household and industrial products in use today, its toxicity has been recognized for many years. Now pentachlorophenol has become controversial in Canada. A labour union is demanding a ban of the chemical, especially of its industrial uses. Some government officials predict this campaign will succeed, as pentachlorophenol was banned in Sweden and Japan. These same officials maintain that the chemical can be used safely, and that there is no need for a ban. They believe that a ban would reflect pentachlorophenol's status as a political chemical.

Perhaps it was inevitable that pentachlorophenol would become a political chemical, as we too believe has happened. Dioxins are produced in the manufacture of pentachlorophenol and other chlorophenols, which are used in the making of 2,4,5-T and Agent Orange. The controversy about Agent Orange – and about the conduct of the Vietnam war – has spread to pentachlorophenol. The regulatory status of pentachlorophenol is not easily resolved because so many issues are involved that are not directly related to it. One such issue is the jurisdictional complexity of regulating pentachlorophenol.

This point requires further explanation. Pentachlorophenol can also be classified as both a pesticide and an industrial contaminant. The rules and regulatory process for pesticides and for industrial contaminants are different in Canada, as they are in most jurisdictions. Pesticides are toxic substances used for the deliberate purposes of controlling insects, fungi and unwanted vegetation, and as such, they are subject to registration by Agriculture Canada. Industrial contaminants are the unwanted, and often toxic by-products of industrial or chemical processes. The assessment of industrial contaminants is usually done by federal authorities other than Agriculture Canada. Standards and their enforcement are a provincial responsibility, and provincial departments of the Environment and Labour may be involved as well as the provincial Workers' Compensation Boards. In other words, because pentachlorophenol is a pesticide used in an industrial setting, several different government bodies, each with a different mandate, are involved in its approval or control. If one of the factors creating political chemicals is jurisdictional complexity, pentachlorophenol is a natural candidate.

In the case of pentachlorophenol, science itself – the actual content of scientific studies – has become the subject of debate. A group of Swedish scientists were subject to some pretty rough treatment in the hands of several regulatory bodies, although their studies are published and often cited in the academic literature. Questions have been raised about the use of science as an instrument of advocate politics. These questions are puzzling because the Swedish scientists do not see themselves as advocates. The pentachlorophenol case is interesting for the study of mandated science, then, because it highlights the relationships between politics and science, and between mandated science and the conventional understanding of academic science.

## Some Background Information

When we speak about pentachlorophenol, we are referring to several chemicals. The products used in wood processing and preservation contain varying proportions of pentachlorophenol, of tetrachlorophenol and of some other chlorophenols. As well, other chlorophenols can be produced as contaminants in the production of commercial pentachlorophenol. The term "pentachlorophenol" is essentially a generic term. It refers to all of the products containing pentachlorophenol used to preserve or process wood.

The term "pentachlorophenol" can also refer to an oil-based solution containing pentachlorophenol. In this instance, "pentachlorophenol" means the chemical used in home repairs for protecting wood on decks or fences from rot, in the forest industry again for preserving wood from rot, and in some countries for tanning leather. "Pentachlorophenate" refers to a product that is usually sold in the form of a water soluble salt. Pentachlorophenate is used for the short-term treatment of wood to prevent staining and discolouration ("anti-sapstain"), and is primarily an industrial chemical. In this book, and in general discussion, the generic term "pentachlorophenol" is used to refer to both of these products, unless it is important to distinguish between them.

In industry, pentachlorophenol, the oil-based solution, is applied to wood under pressure. The application takes place in a closed vessel, and workers do not come into contact with the chemical during its application. They are exposed to pentachlorophenol when they clean equipment, and when treated wood is being stored. Pentachlorophenate, the water soluable salt, can be applied by a variety of different methods, including with paint brushes or a garden hose, by submersion of lumber in a trough, by submersion of the lumber in open "dip tanks", and by spraying wood in enclosed, but unsealed boxes. All but the first of these methods are still in use in industry today, although enclosed spray boxes are now the most frequent method of application. All of these methods result in human exposure to pentachlorophenate.

In North America, about ninety per cent of all pentachlorophenol is used in the forest industry, to treat lumber en route from the mills to the market or to preserve railway ties, telephone poles and any other wood products that come into contact with the ground. Thus, although pentachlorophenol is technically classified as a pesticide, its main entry point into the environment is through industrial operations.

The general public does come into contact with pentachlorophenol. For example, one report stated that: "Preliminary results of a national survey (1978) in the United States indicate(d) the presence of PCP (pentachlorophenol) in the human urine samples analysed at a mean concentration of 6 ppb".[1] People come into contact with pentachlorophenol by using treated wood, and through impurities in the water supply caused by "run offs" from treated wood. Pentachlorophenol gets into the food supply by the use of treated wood in farm buildings and in the shavings on the floor of chicken coops.

ACGIH documentation refers to the dangers of pentachlorophenol in the following manner. "The world literature reveals about 51 cases of PCP poisoning from its use as a herbicide, molluscicide or wood preservative of which 30 out of 51 resulted in death." Health problems from pentachlorophonol exposure are described by ACGIH as, "The survivors of PCP intoxication suffer with impairments in autonomic function, circulation, visual damage and an acute type of scotoma."[2] There is some danger from pentachlorophenol inhalation, but pentachlorophenol is absorbed mainly through the skin, where the result is often chloracne, a long-term irritation. Finally, some scientists claim pentachlorophenol exposure can result in soft tissue sarcomas, a relatively rare form of cancer, Hodgkin's disease and non-Hodgkin's lymphoma.

It is difficult to assess the dangers of pentachlorophenol because two known contaminants are produced in the manufacturing process.[3] These contaminants are "dioxins" and "furans", both of which are classes of compounds. In fact, there are seventy-five dioxins, and several of them are by-products in the manufacturing of pentachlorophenol. The specific dioxin contaminants of pentachlorophenol do not include the so-called deadly dioxin, 2,3,7,8-TCDD.[4] Both the dioxins and the furans in pentachlorophenol should be considered as possible health hazards. The possibility cannot be rejected that the contaminants in commercial pentachlorophenol could be as dangerous as 2,3,7,8-TCDD or that they could be converted to other equally deadly contaminants through degradation or combustion. Moreover, the widespread use of pentachlorophenol as an industrial chemical increases the load of dioxins in the environment, a fact causing concern in the Canadian Departments of Environment and Health and Welfare.

The manner in which the toxic effects in humans develop as a result of pentachlorophenol exposure is not yet fully understood. Nor can the dangers be gauged accurately, because of the association of pentachlorophenol with its contaminants. It is not known yet whether problems are caused by exposure to pentachlorophenol, or to its dioxin and furan contaminants. Although pure pentachlorophenol is both embryolethal and embryotoxic, it is not fully understood how pentachlorophenol acts within the human body to produce its toxic effects. The hazard warnings on the cans of pentachlorophenol are necessary, but the full extent of the hazard remains to be determined.

Environmental damage from pentachlorophenol is the result of chemical spills or "run off" from treated lumber that is being stored. Pentachlorophenol can persist in warm moist soils for periods of up to twelve months, and its extreme toxicity to fish is well documented. The problems can be quite serious, as such spills have resulted in large fish kills.[5] Since fishing is a major industry in British Columbia, pentachlorophenol spills are of considerable economic concern to the province.

It is possible to produce a purer pentachlorophenol, and to exclude most of the dioxin and furan contaminants that are now part of the commercial product. The production of the contaminants is currently an intrinsic part of the production of pentachlorophenol. The cleaner product is only less contaminated because the manufacturer, not the consumer, is left with the problem of dealing with the contaminants. In dealing with pentachlorophenol as a waste product, it is possible to dispose of it and its contaminants by combustion, however dioxins are destroyed only at temperatures of at least 1100° C. There is evidence that dioxins are not destroyed at lower temperatures, but indeed that more dioxins are produced.

Industry has considered producing a cleaner pentachlorophenol. The result was described to us as follows:

> In April 1978, Monsanto closed their plant in Illinois...In October, Dow closed. They were suddenly right out of the market. (The situation was one of) marginal profitability. Competition was very strong. (There were) a lot of environmental concerns. Monsanto got out for this reason. They had a new herbicide coming out which was doing very well. Dow had a clean PCP which Dow wanted the country to use. They had the capacity in Midland for an incinerator to clean up and destroy the dioxins. Nobody would buy the cleaner pentachlorophenol. It was more expensive, so Dow got out of the market. So in Canada, Uniroyal, and in the US, Reichold and Vulcan - they took up the slack.

If a ban was proposed in Canada, a socio-economic impact assessment would be required by the federal government. A government official suggested to us that , "industry would scream blue murder", were such a ban to be instituted. He predicted that if such a socio-economic assessment was done, the economic consequences of banning pentachlorophenol would offset the risk, and the result would be continued use of pentachlorophenol.

It is often argued that, in spite of its dangers, pentachlorophenol can be used with relative safety if proper precautions are used. In the period we studied – 1975 through 1985 – we saw ample evidence of improper handling of pentachlorophenol in the wood processing plants in British Columbia. We have photos of the chemical dripping from leaky facilities. The situation seems to be improving, but many dangerous situations still exist, especially with respect to dip tanks that are used for immersing wood in a pentachlorophenate solution. Whether the change will be sufficient to protect the health of workers exposed to the chemical, and to prevent environmental damage, is a matter of some controversy.

Pentachlorophenol, the oil-based solution, is unnecessary in the treatment of wood if the wood is kiln dried – a more expensive method of treating wood. It is also possible to use other chemicals to prevent rot, but these other materials may be as dangerous as pentachlorophenol. Pentachlorophenate is used mainly for cosmetic and economic reasons. Failure to treat wood with pentachlorophenate does not significantly affect its rate of deterioration. Nonetheless, both industry and government officials claim that it is impossible to sell lumber that is covered with sapstain, and that, as a consequence, pentachlorophenate is considered essential to the economic well-being of the forest industry. Alternatives to pentachlorophenate treatments are also available, but they are also said to be costly.

In general, pentachlorophenol is so widely used because it has been the cheapest alternative available. Until recently, little research has been done, and federal officials record little activity to prepare for the registration of an alternative chemical treatment. A simple description of the chemical tells the story:

> Pentachlorophenols (PCP) and its sodium salt (Na-PCP) are probably the most versatile pesticides now in use in the United States. In fact, collectively, they are the second heaviest used pesticides in the country. They are properly called biocides because they are lethal to a wide variety of living organisms, both plant and animal. PCP is registered by the U.S. Environmental Protection Agency (EPA) as an insecticide (termicide), fungicide, herbicide, algicide, disinfectant, and as an ingredient of antifouling paint. This versatility is due in large part to the solubility of the pentachlorophenols in both organic solvents (PCP) and water (Na-PCP). Thus PCP can be applied to such diverse materials as agricultural seeds (for non-food uses), leather, masonry, wood, cooling tower water, rope and paper mill systems.[6]

Until the last few years, pentachlorophenol was the second most widely used pesticide in the United States. Recently, pentachlorophenol use has declined, and at the time when this book was written, only one company was producing pentachlorophenol in North America.

The oil-based pentachlorophenol is still very heavily used in Canada, particularly in the forest industry on the west coast of the country. There are approximately fifty plants for pressure treating wood with pentachlorophenol, and the annual value of the pressure treated wood products is one hundred million dollars. There are five hundred sites at which anti-sapstain pentachlorophenate treatments are carried out, and the value of pentachlorophenate-treated lumber is more than one billion dollars. For the last few years, pentachlorophenol has been formulated but not manufactured in Canada, but a new manufacturing plant has been proposed for one of the western provinces. This proposal met with a citizen protest when it was suggested that the plant be located in British Columbia.

In Canada, products containing pentachlorophenol, the oil-based solution, can also be purchased by consumers at most hardware and paint stores. Pentachlorophnol is contained in many products in widespread use, such as fence posts and paint. Wood shavings, possibly containing pentachlorophenol, are burned to produce energy in pulp mills. Lumber treated with pentachlorophenol is not supposed to be used in home or farm construction, but there is no requirement in Canada to label the treated lumber, and there is no mandatory accounting system to keep track of large lots of treated lumber. Thus there is no way of knowing how much treated lumber is used in home and farm construction. It is quite likely that consumers and farmers are exposed to treated wood without their knowledge.

## The Standards

The British Columbia standard for pentachlorophenol was adopted without modification from the ACGIH TLV. The documentation to support ACGIH's pentachlorophenol standard is among the least impressive of its products, since it is obviously out of date. References from 1939 are included, and the most recent study cited is from 1971.[7]

Although several government departments are involved in inspection and in dealing with spills of pentachlorophenol, the governmental body with the responsibility for occupational health and safety standards in British Columbia is the Workers' Compensation Board. This Board operates an industry-financed, state-run insurance system dealing with industrial accidents, and industry-related diseases. Employer contributions are used to finance the compensation awards, and the board assesses an industry's fees partly in response to the number of compensation claims. The Board uses standards to determine whether compensation is required. It also conducts inspections of worksites to ensure that its standards are being implemented. The Board has held several hearings in connection with its standards and regulations, but has yet to review the standard for pentachlorophenol.

At a national level in Canada, assessment of pentachlorophenol has taken place through the more or less normal procedures of government. As a pesticide, pentachlorophenol is registered by Agriculture Canada. In addition, significant resources – by Canadian standards – have been allocated to its assessment as a possible contaminant. Canada has commissioned a task force on chlorophenate use in British Columbia, and a Federal Expert Advisory Committee on the dioxins. Quebec has held a public hearing on pentachlorophenol. A group of citizens have launched a civil law suit to stop the spraying of their land with 2,4,5-T, raising issues related to the safety of pentachlorophenol. A number of studies have been done, mainly by federal government departments or through these task forces and committees.

These studies have been used to promote a clean up of industrial worksites, but they have not resulted in a ban of pentachlorophenol. At the time of this writing, they have also not resulted in a change in the British Columbia Workers' Compensation Board pentachlorophenol standard. In fact, there is little direct contact between the federal authorities that conduct assessments, and the provincial Compensation Board that sets standards. An official with a provincial research council described the relationship in the following terms:

> We talk to them (WCB). We are in technical communication with them from time to time, but we don't have input into their decisions, other than we have been invited when they were going to review admissable concentrations and various other aspects of the regulations *(editorial note:* in 1972 and 1976; pentachlorophenol was not discussed in these reviews.) We were invited to put someone on that review committee. That review committee hasn't, in fact, met (at the time of the interview), because of the disorientation of the WCB's philosophy. We don't have any association with the WCB formally or any other way other than from individual contacts to exchange technical information.

Both Agriculture Canada and the provincial authorities in British Columbia could impose a ban on pentachlorophenol, either by doing so directly or by making the standard so stringent that it would be tantamount to a ban. Canada has restricted some of its industrial uses, but efforts have mainly been to institute codes of good practice.

Groups seeking a ban on the use of pentachlorophenol cite the examples of Sweden and Japan. Sweden banned pentachlorophenol in 1978. Because much of the wood produced in Sweden is kiln dried, or is distributed to markets nearby, the economic consequences of a ban on pentachlorophenol in Sweden are considerably less than would be the case in Canada. The Swedish decision appears not to have been directly linked to the dangers of pentachlorophenol use in the forest industry or to findings of the Swedish scientists, who linked the use of pentachlorophenol in the forest industry with health hazards. The Swedish decision is commonly attributed to policies of dioxin control and the controversy over Agent Orange. The Japanese ban on pentachlorophenol was apparently due to contamination of fish, a dietary staple in Japan.

The United States has been reassessing the registration status of pentachlorophenol for several years. It has now banned pentachlorophenol as a consumer product and imposed some restrictions on its industrial uses. Meanwhile, one of the last remaining American manufacturers of pentachlorophenol, Reichhold Chemicals, has ceased production of pentachlorophenol, arguing that the costs and risks are too high.

## History of the Controversy

It is necessary to relate several different versions of the pentachlorophenol controversy, because there is little consistency in the histories of the controversy provided to us by union members, industry, federal and provincial government officials and academic scientists. For example, the history of Canadian government involvement is different from the story provided by the union. The events in British Columbia touch upon the scientific debate in regulatory tribunals outside Canada only peripherally. There is a discrepancy between how the scientific studies conducted by the Swedish research group are viewed in the academic and regulatory communities. The Canadian governmental assessments of pentachlorophenol included the Swedish studies, but the Swedish studies played a minimal role in the controversy in British Columbia. We have chosen to recount three different versions of the controversy, the local story, the regulatory chronology and the regulatory controversy outside Canada.

### *a) The Local Story*

The effect of pentachlorophenol on workers was first raised by a labour union in a small community in British Columbia. The union had two sources of information. First, in 1976, two local provincial officials gave scientific papers on pentachlorophenol at a meeting of wood manufacturers in Portland, Oregon. Although they argued that the dangers of pentachlorophenol were not very significant, the union took their interest as a sign that concern was warranted. Second, the union had access to an internal memo that had been leaked from one of the pentachlorophenol manufacturers. The memo discussed the link between pentachlorophenol and dioxins.

The union scheduled seminars on health and safety issues, focussing on pentachlorophenol. A union official told the story as follows:

> Our people generally began asking more and more questions. (One person in one of the locals)...took up the issue as his thing. As he became concerned, he sent for all the information and started to work with a committee. They put out a newsletter to the employees within the operation, generally advising them about the chemical. And they took it up in a committee with the employer, because there were serious problems in the mill.

One such seminar, in Seattle in 1980, was particularly influential. There, union members received information produced by NIOSH about chlorophenol. "It wasn't specific," one delegate told us, "and it simply mentioned most of the acute effects, but it did mention the possible long term liver and kidney damage." At the same seminar, they also heard from one of the Swedish scientists, who had claimed that pentachlorophenol exposure was dangerous. The union had never doubted that its workers were being exposed to pentachlorophenol, and now it believed that workers had suffered from that exposure. It came to believe that the dangers of pentachlorophenol had been underestimated by regulators.

A union spokesman described the situation in the mills at the time, as follows:

> They had sprays and they leaked. I gather that the sprays were plugged, so the spray wasn't directed in one direction. It was going all over the place. They asked the maintenance workers to deal with it. The maintenance workers wanted protective clothing, but the company wasn't willing to do it. The maintenance people refused to work on it. The employees were then working with it (the chemical) on the floor, and it was wet on the floor...They were getting it on their jeans. They didn't have proper gloves; they had leather gloves that soaked through. Guys had dermatitis down their legs and on their hands. So they tried to do something about it...I don't think the company said "no" at first... It was just delayed, delayed, delayed.

Union members returned to Canada from the Seattle seminar, and demanded a cleanup of a local mill that was using pentachlorophenate. Even a company official described the situation at the time as "a horror show". The company believed, however, that any action on pentachlorophenol would threaten its continued use for treating wood. Apparently nothing was done in response to the workers' complaints.

Faced with inaction, the union members sought the assistance of the provincial Workers' Compensation Board. The WCB inspected the mill and wrote orders for a clean up of the facilities. Still, the cleanup was not undertaken. In December 1980, the workers went out on a wildcat strike, indicating that they would not return to work unless conditions improved. The workers believed that they acted legally by refusing to work under unsafe conditions, but they failed to follow the proper procedure under provincial legislation for such actions.

A union spokesman described what happened next:

> The Compensation Board came in and wrote orders on basic clean up... There had been violations before, but they'd never written orders or made sure they were complied with. This is what the committee told us at the time. The mill said, 'All right, we'll comply with the orders but you have to return to work.' The union said, 'We'll not return to work until all the orders are complied with', plus they added another four point demand which included a health study of all the members. Interestingly enough, in response to this, the company managed to comply within 48 hours.

Labour relations returned to normal when the cleanup work was undertaken and a proposal initiated for study of workers' health problems. The union assumed that the company would contribute half of the cost of the study, because its members had returned to work on the condition that research be undertaken.

The level of company participation in this and further health research was a continuing source of union-management conflict.

The union decided to proceed with the research, even without assurances of company financial support. A union spokesman said, "There was just not the money to do a full epidemiological survey ... so we did it for the twenty thousand dollars." After consulting with management about who should conduct the study, the union commissioned it from an industrial health group located at, but independent of one of the local universities. The study was later presented at an AIHA conference and eventually published in a scientific journal.[8]

The company had decided in the meantime that it was facing an industry-wide problem, and that industry-wide support was required. This support came from two different sources. It came from the Canadian Wood Preservers Association (CWPA) which had been established in 1979, partly to develop standards. CWPA's membership was mainly drawn from industry, but government officials participated. A review of the available papers from the first two annual meetings of CWPA suggests that CWPA would have been a willing ally. Many of its members saw pentachlorophenol as highly beneficial and the task of the organization as allaying concerns about its use. One corporate member suggested in one meeting, for example, that there was "no solid evidence of hazard" and that "good practice (wa)s the only real concern". An official from B.C. Hydro, a government owned utility, argued at the same meeting that there were no serious environmental or work-related concerns with pentachlorophenol, and that "CWPA must ensure that accurate and factual rebuttals are prepared to defend our use of chemicals."

Second, the company approached the Forest Industry Industrial Health Research Program (FIIHRP) to carry out research. FIIHRP had been established in 1979, partly as a result of a study undertaken on behalf of the Council of Forest Industries on the use of hazardous chemicals, and it had already done several small studies relating to pentachlorophenol. In response to the company's request, FIIHRP asked B.C. Research, an independent research consulting company, to do a new study relating to pentachlorophenol.[9]

Now there were two new studies, the union's and the one commissioned by FIIHRP. The difference between them was significant. The union commissioned epidemiological research, seeking evidence that the chemical was dangerous to exposed workers, and using a worker questionnaire to identify their complaints. The union study focussed on symptoms of discomfort and on health problems. In contrast, the mandate for the FIIHRP sponsored research was simply:

> to come up with a recommendation or series of recommendations on what kinds of studies would be feasible to obtain useful results on the health effects of chlorophenates on saw mill workers of the province.

Workers were not questioned in the FIIHRP sponsored study, nor was an effort made to quantify either exposure to pentachlorophenol or any harm resulting from that exposure. The FIIHRP study was simply to determine whether an epidemiolgical study was feasible.

A FIIHRP steering committee was set up to oversee the feasibility study. The union participated in this FIIHRP steering committee on the condition that its own study would also be reviewed. In theory, the feasibility study should have complemented the epidemiological research commissioned by the union, because the feasibility study was posited on the assumption that more comprehensive epidemiological research would be conducted if adequate data were found to be available. As one FIIHRP committee member told us:

> All that had been done at the time were, I think, three site specific studies. (one was the union sponsored study). (The researchers for) two of the studies conceded that the number of participants was really too small to draw any statistically reliable conclusions. So the industry... at the urging of (one of its employees) who is an epidemiologist decided to retain...an epidemiologist to carry out a study to see whether...there was enough information around to do a long-term health study of whatever kind.

In practice, the feasibility study gave rise to controversy. The situation was described to us, as follows:

> At that time in 1980, the committee – I should say the industry – no!, everyone was scrambling around trying to find out what we are going to do. Suddenly the pot boiled over. The union said ban chlorophenates; we aren't going to use them any more; get rid of them. The industry was taken by surprise and realized they knew little about it. There was a good deal of grasping in thin air to try to find out whatever they could about chlorophenates. We did studies which I think were almost, at that time, shotgun affairs in that they wanted quick results and quick information... We were trying to get some handle on what was the exposure and what are the problems. We didn't have any health studies to contribute to that... The committee was extraordinarily polarized at the time.

There was no basis for trust among the steering committee participants, and neither research nor further discussion provided one. One member told us the story:

> I don't know whether it was the first or the second meeting...every meeting started off in this way. We had about an hour of ventilating spleen and invective from labour and management before the bubble deflated and we were able to discuss things in a rational way. I recall saying in one of the early meetings to the committee: do you want to have a health study or don't you? If the answer is no, we'll go home and pack it up.

Nonetheless, as a result of the feasibility study there was an agreement in the steering committee to do another epidemiological study, and its terms of reference were finally agreed upon after eighteeen months of acrimonious discussion.

The FIIHRP committee then sent out five requests for proposals to conduct the work. The request was sent to the Swedish scientists, whose research had initially raised questions about the safety of pentachlorophenol for the union, but the Swedish scientists chose not to submit a proposal. Two other proposals were received. One of these proposals was from a controversial and internationally known researcher on cancer, whose work had been linked to the identification of numerous cancer causing hazards. The union favoured this proposal, because the person involved was "high profile enough to command respect from everyone." It was, however, twice as costly as the other proposal, which came from one of the researchers who had conducted the original FIIHRP feasibility study. The second proposal was supported by the industry representatives on the FIIHRP committee, although the scientist involved was not, formally or in any evident sense, an industry scientist. According to a union advisor, industry's position was that they wanted a Canadian to do the study.

The steering committee was now in a deadlock. To resolve it, with the union's reluctant consent, the committee decided to send the two proposals out for scientific peer review. It chose five scientists to conduct the review. It received one moderately positive evaluation for each of the two proposals. Three other reviewers had serious criticisms of both proposals. Although all of the reviews were at least somewhat critical, the nature of their criticisms differed. Scientists could not, it seemed, resolve the issues any better than the interest groups participating in FIIHRP. Nor did the scientists reach a consensus. In the short term, FIIHRP decided to do nothing and indeed, the steering committee ceased meeting shortly thereafter.

The university scientist who had conducted the feasibility study decided to proceed without FIIHRP or its steering committee. He submitted his proposal for a mortality study – or, rather, a revised version of it – to the federal Department of Health and Welfare and requested funding. As was required, he indicated that he had consulted the union and that the union supported the research. The union disputed his claim. An official told us:

> The union wanted to do it too (The second epidemiology study). See, the study that was submitted to Health and Welfare was a different one than had been submitted to FIIHRP. The one to Health and Welfare - no one knew about it... We were given a copy at the last meeting of the committee before the union and the rest of us said, "to hell with FIIHRP". We were given the study proposal as information and told that he's going to get $330,000 to do this.
>
> We said, "what!" And we looked inside, and in the application form it said that the study had been fully discussed with the union involved. The union had never seen this proposal....It became a kind of political game. How does it look, ... for the union to be seen as obstructing research...How are we going to respond to this? We said, "Fine, let's do a mortality study ... but only on the condition that you start the work by instituting a temporary ban on the product, as had already been demanded...They should put a greater priority on taking that same $300,000 and

> putting it into applied research into finding and testing alternatives." And that was our response.

The researcher has responded:

> We put in our proposal and then heard vaguely that the (FIIHRP) committee had met a couple more times and wasn't getting anywhere. It turns out the thing was just stuck. And we nearly got the (federally funded) study completely screwed up because the union got the feeling that, perhaps correctly, I don't know, that the companies were trying to stall...The unions were very keen on (the other proposal), as he's a sort of friend of the unions. But the chairman of the committee was saying that (it couldn't) accept it (the other proposal) because it cost too much money and (the proposed research was) not doing what the committee asked.
>
> And so everything came to a standstill with the chairman saying in the last meeting, "in any case, if (scientist X) is going to do the mortality study and get funding from (the federal government), then there is no need for our study". The unions took this as being an excuse for not doing anything, and the next thing was that I got a letter saying that the union did not support it.

The union was not only concerned about the lack of consultation. It also had questions about the design of the proposed research. Echoing concerns we also heard from some academic scientists, a union activist described the situation, as follows:

> We were concerned about the power of it (the study under consideration for federal funding) in terms of finding significant increases for certain types of mortalities due to certain types of cancers. It was questionable. The elevated risks that showed up in the Swedish work for certain types of cancers was such that there was an extremely good chance that this study wouldn't be able to determine their significance...because of the small numbers of people they would use. They were rare cancers.
>
> They were going to do a retrospective mortality study.... The Swedish study is all case-referent, or case-control studies. There has already been a long debate in the international research communities as to which (approach) is more sensitive to what...I think any respected research epidemiologist would say that if you are looking for rare cancers, you need a case control method. ...(The proposal's author) said that there might be a problem there, and therefore that they would have to wait until there were further deaths recorded, another decade.
>
> And he also said in his justification to Health and Welfare that, in terms of the rationale ... that in the case that they didn't find an excess (harm among those exposed), it would allow the regulatory community (industry and government) to assuage fears among the public and to perhaps review recent restrictions made on that chemical. ...This one study (with no acknowledgement of its limitations of other studies) is going to allow the Canadian government to deregulate.

An exchange of letters took place between the union and the Minister of Health and Welfare, citing the union's complaints about the proposed research.

The union said that the research would be (and that it had been) used to avoid banning pentachlorophenol, and to deregulate some aspects of workers' safety. It questioned the way that exposure would be gauged for experienced workers: it saw the proposed use of a "typical day" as questionable on methodological grounds. It was concerned that the proposed research would exclude some of the most exposed workers – namely, those who were students or who were hired in the summer to do the "dirty work" in the mills. It was concerned because the most serious of the potential health effects of pentachlorophenol are relatively rare forms of cancer that might not have shown up in the type of study being proposed.

The union was also worried because other epidemiological studies – its own and those of the Swedish scientists – which did show a positive link between pentachlorophenol and human harm – would be disregarded. The union felt that the new study might be used as a reason to remove restrictions on pentachlorophenol, and it drew attention to the lack of alternatives to pentachlorophenol being developed.

The union withdrew from FIIHRP in 1983. The union stated its position in a letter to the Forest Labour Relations Council, as follows:

> "In all the years that the FIIHRP program has existed, over a million dollars has been spent by the industry, and yet it is doubtful that the health protection of even one worker has been improved."[11]

It also stated:

> If we couldn't have equal representation on FIIHRP and some more say on some of the research being done by groups that were chosen by the committee, then it wasn't worthwhile.

Along with industry and government officials, the union had also been involved, in 1981-2, with a Federal-Provincial Task Force on the development of a code of good practice for handling chlorophenates in wood processing. After the union withdrew from FIIHRP, it refused to participate in the development of a second code of practice for pentachlorophenol, the oil-based solution. The union has argued that the codes were and are being used to forestall other actions with respect to the chlorophenols. About this, the union was undoubtedly correct in its perception. The union has also suggested that the codes were also being used to promote the idea of *new* and expanded uses of the chemicals.

A slump in the forest industry has diminished the union's resources for dealing with health and safety issues. In spite of this, the union is still actively pursuing its claim that pentachlorophenol is dangerous and that it should be banned.

### *b) the Regulatory Chronology*

Pentachlorophenol was first registered as a wood preservation product for use in

Canada in 1941. The original registration of pentachlorophenol by Agriculture Canada was on the basis – as it was for many pesticides – of data that would not withstand scrutiny today. It simply indicated that the pesticide would be effective for its stated use and that pentachlorophenol was probably not toxic to its applicator if proper precautions were used.

In 1971, Agriculture Canada recommended – it had no power to ban the chemical for specific uses at that time – that pentachlorophenol not be used for tanning and preserving hides, but no studies were released to support this decision. In 1970-71, under its new legislation, Agriculture Canada gave more attention to pentachlorophenol. The first assessment in 1970-71 was a simple one. In 1974, a sampling program – methodology to conduct it became available only in 1973 – was undertaken by Agriculture Canada to investigate the dioxin content of pentachlorophenol, searching mainly for the dioxin, 2,3,7,8-TCDD. None was found, but other isomers were not fully investigated, and the regulatory status of pentachlorophenol remained intact.

In 1976, the regulatory situation became much more complicated. An *Environmental Contaminants Act* was enacted in conjunction with the newly created federal government department, Environment Canada. Under the powers granted in the *Environmental Contaminants Act*, Environment Canada could request information about chemicals to determine their priority for environmental assessment. Listing a chemical as a priority chemical did not, in itself, mean that new restrictions would be introduced, or even that a significant hazard had been identified. Nonetheless, chemicals on the priority list were likely to receive significant regulatory attention and to be subject to relatively comprehensive assessments. As one Environment Canada official told us:

> Anyone that has information on any of the substances is invited to communicate that information...The list is not indicating the start up of an irreversible process that is closing on the company or the substance or is going to affect the companies that deal with these substances. It is just an indication of a particular interest for these substances... except where regulations are under development. What you want to establish about a priority chemical...is there a significant danger?...It's a kind of demonstration process you have to go through to convince people that what you want to do is based not strictly upon being overly careful about something.

Several departments, including Agriculture Canada, and the Department of Health and Welfare, included pentachlorophenol on their list of proposed priority chemicals. In 1977, Environment Canada responded to these requests by making pentachlorophenol one of its first priorities. A number of studies related to pentachlorophenol were then conducted by consulting firms under contract to the federal Environmental Protection Service.

Under the *Environmental Contaminants Act,* companies are not compelled to release information. Priority listing of a chemical by Environment Canada does not necessarily mean that further information about its toxicity would be made available, either to Environment Canada or to the public. Environment Canada does not have responsibility for the regulation of pentachlorophenol, as long as it remains a pesticide under the jurisdiction of Agriculture Canada.[12] Agriculture Canada can request information if a chemical is a registered pesticide, but Agriculture Canada respects the companies' desire for confidentiality.

Shortly after Environment Canada listed pentachlorophenol as a priority chemical, it was scheduled by Agriculture Canada for reevaluation. Then, in 1977, several accidents occurred, in Canada and elsewhere, involving pentachlorophenol. These accidents increased the profile of pentachlorophenol as a potentially hazardous chemical. A product containing pentachlorophenol – which was being used illegally for treating leather – was withdrawn from use by Agriculture Canada. Later the same year, Environment Canada began its own technical review of the chlorophenols.

In 1978, two pentachlorophenol manufacturing plants were closed, an investigation of the "off flavour" musty odour of chickens was undertaken by Agriculture Canada (pentachlorophenol was found in the wood shavings used for bedding the chickens), a spill of pentachlorophenol was reported in British Columbia and the American environmental regulatory agency began its review of the status of pentachlorophenol. In 1979, a memo was issued by Agriculture Canada to the manufacturers of pentachlorophenol stating that the status of the chemical was under review in Canada. Some of its specific uses were suspended. The registrants of the chemical were notified that it was being reevaluated. Late in 1979, Agriculture Canada suspended additional uses of pentachlorophenol and proposed new cautionary labelling for products. In 1980, two Canadian accidents involving pentachlorophenol occurred, and two manufacturing plants – one in Canada and one in the United States – were closed. In British Columbia, a Federal-Provincial Task Force was established, to work on the first code of good practice. A further notice of changes to the regulatory status of pentachlorophenol was released. In 1980, as well, the Department of Health and Welfare took regulatory action, stating that food containing chlorinated dibenzo-p-dioxins – this includes contamination with pentachlorophenol – should be considered contaminated.

Several research reports were published between 1980 and 1983. The final draft of an Environment Canada document on chlorophenols was made available to a "contaminants committee", which included officials from both Environment Canada and the Department of Health and Welfare.[13] The Environment Canada report stated that pentachlorophenol was ubiquitous in the Canadian environment, that some food and feeds had been contaminated by chlorophenol during their storage, and that the chlorophenols were of toxicological significance especially for aquatic organisms. The report noted that "experimental results also indicate a direct relationship between the presence of tumoregenic lesions in test animals and the isomeric structure of chlorophenol."[14] The contaminants committee recommended that additional information on the commercial flow of chlorophenols in Canada should be obtained, using the notice provisions of the *Environmental Contaminants Act.*

The National Research Council issued two assessments of the dangers of contaminants associated with pentachlorophenol. The first of these reports concluded that "epidemiology and detailed exposure studies should be carried out with identified high exposure groups".[15] It recommended that all Canadian sources of furans with the potential for relatively high human exposure should be identified. The second report recommended the continued monitoring and assessment of the dioxins already in the environment. It also recommended that research be undertaken to determine the sensitivity of people to chlorophenol.[16]

Environment Canada and Health and Welfare Canada convened an expert advisory committee on dioxins and their health-related hazards. Its report drew particular attention to dioxin contamination that was spread through the use of contaminated wood shavings. The report stated that, "the major area where there is still concern is the use of pentachlorophenol and tetrachlorophenol in wood preservation and protection."[17] Chlorophenols were identified as the second most important source of dioxins in the environment, and as the most important pathway of human exposure to dioxins. The joint report recommended changes in the manufacture of pentachlorophenol to reduce the levels of dioxins, a program of environmental monitoring, and further research to determine the specific levels and types of dioxins in pentachlorophenol. It called for the elimination – where possible – of the sources of environmental dioxins. This report was distributed to the Toxic Chemicals Committee, which was made up of representatives from five federal government departments.

What is the regulatory situation today? One federal official described it as follows:

> Chlorophenols do not need and will not be regulated under the *Environmental Contaminants Act.* It is residual legislation. There is adequate power under the *Pest Control Products Act* (under which Agriculture Canada is the registration authority) to regulate chlorophenols.
>
> They (Agriculture Canada) were quite prompt in putting in those regulations at the start of 1982...on restricted uses of chlorophenols. Those regulations went a long way to removing the substantial hazards from chlorophenols getting into chickens and poultry on farms and in a lot of homes in decks, etc.
>
> (These changes) were negotiated with us (Environment Canada) and with an official in the occupational health group in Health and Welfare... They are now working on...looking at new suppliers of chlorophenols and looking at dioxin levels in it and we're pushing them very hard to come up with a maximum allowable (standard for) hexa, hepta and octa dioxins in commercial chlorophenols, as they've done with 2,4,D.

The last point indicates a curious inconsistency: Although the dioxin content of 2,4-D (and some other herbicides) has been restricted, it has been estimated that the amount of dioxins entering the Canadian environment from the use of pentachlorophenol is more than 100 times that entering from 2,4-D before 2,4-D was restricted.[18]

We started this chapter by saying that no one disputes the fact that pentachlorophenol is dangerous. Yet there is conflict about the future regulatory status of pentachlorophenol. On one hand, after almost ten years of activity, some government officials tell us that they are not very concerned about pentachlorophenol exposure. To them, the danger seems under control. A provincial official told us, for example:

> We don't really have any evidence to suggest that pentachlorophenol is a serious problem in terms of severity of symptoms. It's a problem in as much as there are so many workers exposed to it. But my personal feeling is that it is probably fairly well controlled, because we don't see the claim for... ill-effects...I think the problems we see are skin problems and we might get a handful of claims regarding that every year.

Is pentachlorophenol not very dangerous after all, in spite of several government reports expressing concern about its safety? No one actually said this, or suggested that pentachlorophenol should be regarded as a safe chemical. Officials told us that the situation in the pulp mills had improved, but none claimed that the hazards were adequately controlled. Nonetheless, we have the sense that most government officials consider the controversy to be resolved.

On the other hand, some federal officials still think that pentachlorophenol will be banned. One expressed his assessment, as follows:

> So make no mistake. From here it may look pretty good to the forest industry, and I think that Agriculture has reason to believe that the stuff can be cleared up by changing the chemistry. But the problem is, even if you clear it up, the molecules can still react to form dioxins, even after it is out there. So there's no solution in the long run. We can cut down the immediate level considerably, but you can never get rid of it ... Plus the fact that the ... contamination hasn't even hit the public consciousness yet, as to how much dioxin they are sitting on. That (dioxin) is there in fish, it is in the vegetation, it is in the soil, it is in the drinking water. See, wait another five to ten years, till all the bad news comes out and it is "goodbye, pentachlorophenol". I mean even if the government didn't want to act, there would be popular sentiment to get rid of it. And quite rightly so. Why should we take a chance if we don't have to?

### *c) The Regulatory Controversy*

In 1977, a group of Swedish scientists published a small, exploratory study on exposure to pesticides, including pentachlorophenol. The study drew upon their previous research and that of other scientists.[19] It examined the records of pesticide exposure of a group of forestry workers, and indicated that exposed workers appeared to have a greater risk of a relatively rare form of cancer. This study was followed two years later by another more comprehensive one by the same researchers. It also suggested that a link could be drawn between pentachlorophenol exposure and the development of soft tissue sarcomas. Several more articles by the same authors[20] have now been published, based on their continuing program of research. These articles affirm their initial conclusion about the possibility of harm from exposure to pentachlorophenol.

The Swedish studies were published in established academic journals, and they have been analysed by other researchers.[21] By and large, the assessments in the academic journals have been very positive, and prestigious scientists have attested to their merit. From an educated layperson's perspective, the Swedish studies are indistinguishable in terms of their quality, and the range of issues they address, from other work in the same field.

The Swedish studies have also been examined extensively by a number of regulators and commisssions. These assessment have been much less positive. Indeed, many of the regulatory assessments have been negative ones. It is worth noting that many of the regulatory assessments have not been directly concerned with pentachlorophenol, but have been focussed on 2,4,5-T, or on Agent Orange and/or dioxins.[22] In such cases, chlorophenols have been of interest because they are precursors of 2,4,5-T, or because 2,4,5-T is a constituent of Agent Orange. The Swedish studies are controversial not only because of the importance of pentachlorophenol to the forest industry, but also because any dangers associated with pentachlorophenol are related to the status of 2,4,5-T and Agent Orange. Directly or indirectly, the assessment of pentachlorophenol is associated with an assessment of the conduct of the Vietnam war and with several major legal issues arising from it.

It is worth reviewing some of the regulatory assessments, and the response of the Swedish scientists to negative assessments of their work. In 1979, for example, the JMPR reviewed the status of 2,4,5-T, combining its assessment of 2,4,5-T with a review of pentachlorophenol. Its 1981 draft toxicology report stated:

> The 1979 JMPR expressed minimal concern with the residues of 2,4,5-T in food, but indicated that studies were required regarding the bio-accumulation of 2,3,7,8-TCDD in mammalian tissues. Concern was also expressed about the TCDD levels in technical 2,4,5-T. Both of these concerns have been resolved and the meeting has therefore estimated an ADI for technical 2,4,5-T that contains not more than 0.01 mg/kg TCDD.
>
> Studies considered desirable by the 1979 Meeting included any available epidemiology studies. The present meeting considered a number of such studies which were negative or had been discredited. The study (Hardell and Sandstrom, 1979) detailed in the monograph of the present meeting was not relevant to 2,4,5-T containing less than 0.01 mg/kg. Data gathering by telephone survey of individuals occupationally exposed to 2,4,5-T was not considered to be a valid method of epidemiology...[23]

The JMPR evaluation was disputed by the Swedish scientists in a letter they sent to its chairman. They pointed out that their data were based on questionnaires, and not just on a "telephone survey". They argued that the telephone data were supported by data from other investigations. Thus, they defended their use of telephone surveys as one of several legitimate (and established) techniques for obtaining epidemiological information. The Swedish scientists claim that they never received an answer from the JMPR.

The Swedish scientists' correspondence with officials in the European Common Market suggests that a similar conflict occurred between the Swedish scientists and the expert committee conducting assessments for the Common Market standards. They claim they received no reply from the chairman of the Common Market committee.

*International Register of Potentially Toxic Chemicals* published an evaluation of 2,4,5-T, suggesting that "no evidence is found that disqualifies 2,4,5-T as a safe herbicide for the environment". When this evaluation of 2,4,5-T was challenged in a letter from the Swedish scientists to the editor of the *Register*, the director of the *Register* responded that the evaluation in the IRTC Bulletin did not reflect the views of IRPTC, but was instead simply a quote from a British Advisory Committee. The IRPTC, he stated, does not evaluate chemicals, but simply takes note of evaluations that have been done by others.

The British Advisory Committee on Pesticides had commissioned an evaluation of the Swedish work in 1983. The evaluator was Robert Kilpatrick, who had previously done a number of reviews of pesticides for the Committee. The Kilpatrick report came to Canada when Kilpatrick appeared as a witness for the defendant, the Nova Scotia Forest Industry, in a trial on 2,4,5-T. The Kilpatrick report suggested that the Swedish studies should be regarded as inaccurate, because of the bias introduced when workers had been asked to recall their exposure to pentachlorophenol. Kilpatrick's evaluation of their work was also disputed by the Swedish scientists. In response to Kilpatrick's criticisms, the Swedish scientists cited their discussions of methodological issues in a number of their other studies. In a letter to the British Minister of Agriculture, Fisheries and Food, they commented:

> it is too simple a method by a committee to refute our findings without considering the methods to avoid recall bias that we have discussed in our studies and in the letter appearing in *The Lancet*.

Again, the Swedish scientists claim that neither Kilpatrick, nor the British Advisory Committee, nor the Minister responsible for it responded to their defence of their work.

The Swedish studies were charged with yet another form of bias in the course of an Australian Royal Commission of Inquiry into the effects of defoliants (Agent Orange) in Vietnam. In this case, a Swedish government official testified that one of the Swedish scientists contacted the press and released some of their preliminary results to the media. This was supposed to have occurred while the scientists were still gathering their data, thus prejudicing the outcome of the study. In a letter to the Australian Commissioner, two of the Swedish scientists have vigorously disputed their antagonist's testimony.

According to their account, the Swedish Environmental Protection Agency, a sponsor of their research, requested that one of the scientists come to Stockholm to present some of their results. When this scientist responded that all of the data had been collected, but that the analysis was not completed, the agency persisted, and the scientist complied. At the session with the environmental agency it was suggested that the scientist should not contact the media. The scientist's explanation of the fact that the suggestion had been made was that some officials did not wish the spraying season to be interrupted by a controversy. He responded by saying that the scientists would not initiate contact with the media, but that there was a moral obligation to release documented information if asked to do so. Subsequently, the press did contact the Swedish scientists, and some of their results were published in the popular press. In their letter to the Australian Commission, the Swedish scientists claimed that the accusations of bias were totally unfounded, since all data collection had been completed well before anything was reported in the press.

Accusations of bias were only the beginning of the Australian Commission's negative evaluation of the Swedish studies. The studies were found to be lacking in substance. It was stated in the Commission's final report that, "Dr Hardell *is careful at least* about final conclusions" (emphasis added). The report commented further, "Professor Axelson took *no responsibility for the accuracy* of the exposure data", and "Later he sought to *escape the consequences* of this by seeking...".[24] Another example of the tenor of the Commission's evaluation is, as follows:

> ...then the inclusion of the Table in his study would amount to a gross departure from the standards of scientific honesty which one is entitled to expect...such a situation underlines the loose, untrained (sic) and inexperienced approach which was involved in the study and emphasizes the unsatisfactory nature of the exposure data.[25]

The Swedish scientists responded to the Australian Commission's final report by claiming that sections of the final report of the Australian Commission were taken – almost verbatim and without citation – from the submission by Monsanto, the chemical company producing Agent Orange. One of the scientists also provided an extended rebuttal, which stated that "the mistakes and misunderstandings by the Commission regarding my work are serious and numerous and the Commission does not even seem to be able to read and cite published papers in a correct way..." The rebuttal letter from the Swedish scientist stated further:

> (My) opinion is furthermore that the Commission arranged a trap for me by asking me to participate in the hearing thereby being able to misquote me, to interpret my studies in the wrong way, to cite only what is suitable for their purposes.

We are not alone in viewing the Australian Commission's evaluation as unduly harsh, and overly emotional. In an article in the *Medical Journal of Australia* entitled, "Storm in a cup of 2,4,5-T", Bruce Armstrong says:

> ...in true adversarial legal style, the Commission sets out to demolish not only their data but also their credibility. The report frequently described Axelson as "admitting" or "conceding" certain facts, as if unwillingly.[26]

The Commission's report was licence for further attacks on the work of the Swedish scientists. A letter from another epidemiologist in the regulatory controversy suggested that not just the studies, but indeed the scientists themselves should be dismissed. The letter said:

> Your (The Commission's) review of Dr X's work, with the additional evidence obtained directly from him in an interview, shows that many of his published statements were exaggerated or not supportable and that there were many opportunities for bias to have been introduced in the collection of his data. His conclusions cannot be sustained and in my opinion, his work should no longer be cited as scientific evidence.

His letter, which praises the report of the Australian Commission, continued by stating:

> It is clear, too, from your review (The Commission's final report) of the published evidence relating to 2,4-D and 2,4,5-T (the phenoxy herbicides in question) that there is no reason to suppose that they are carcinogenic in laboratory animals and *that even TCDD(dioxin), which has been postulated to be a dangerous contaminant of the herbicides, is at the most, only weakly and inconsistently carcinogenic in laboratory animals.* (emphasis added)

His letter suggests to us that much broader issues were involved in the Australian assessment than the specific content of the Swedish studies. It suggests that his skepticism about the dangers of dioxins generally was influencing his specific evaluation of the Swedish studies of pentachlorophenol. It suggests that at least for this critic of the Swedish scientists' work, pentachlorophenol had become a political chemical.[27]

In each of the regulatory evaluations of the Swedish studies, emphasis was placed upon the limitations of the methodology and upon all the areas of uncertainty in their conclusions. The scientific credibility of their authors was questioned, although not always explicitly. It was argued that the questions about safety raised by the Swedish studies were completely unfounded, and that their discussion of broader safety issues was inappropriate to their role as scientists.

## Discussion

### *The relationship among the three accounts of the controversy*

The participants in the controversy in Canada were certainly aware of the Swedish research. One of the Swedish researchers spoke at the seminar in Seattle. The union viewed the Swedish scientists as natural allies, seeking to engage them for the comprehensive epidemiological work that FIIHRP was planning to undertake. The Swedish scientists did not submit a proposal, however, and we have no evidence of a continuing relationship between the Swedish research group and the union. In effect, the union's main concern was with the experience of British Columbian workers with pentachlorophenol. It would have welcomed *any* epidemiological research that would have documented its experience in a manner that regulators would find convincing.

Officials from the federal departments spoke highly of the Swedish research, yet they did not rely upon it. There is an easily observable difference between the government sponsored research on both dioxins and pentachlorophenol and the Swedish studies. The Swedish studies are academic research, involving an original study of primary data. The government sponsored studies consist of reviews of the scientific literature and of studies conducted – usually by companies – for the purposes of regulation. This is true even of the NRC criteria documents, which are seen by government officials to be academic documents. The chronology of events and studies provided to us by Agriculture Canada takes account of the activities of all the government departments, of plant closures and of accidents, but it does not mention the Swedish studies.

There is a relationship between the union's account of what happened in the pentachlorophenol controversy and that of the Canadian government regulators, even though the chronology provided by Agriculture Canada also does not mention critical events from the union's perspective. From the perspective of the government regulators, the union is seen as a catalyst for the controversy and for a much needed assessment of pentachlorophenol.

In spite of these overlaps, the three accounts of the controversy are very different. This is an important observation for the study of mandated science. When the study of mandated science is approached from the perspective of chemical assessment and regulation, it appears as though a rational process can be designed for evaluating a chemical, its risks and its regulation. In this rational process, the function of assessment can be separated from that of regulation. First scientists are to be consulted, and expert committees or "panels" are used. Then the scientific assessment is to be subject to public hearings. Finally, the debate shifts to the regulatory arena, where scientific assessments are to be complemented by studies of the costs of regulation and the benefits of the chemical in question. The regulatory process often includes some public hearings as well. The result of this process – often called "risk evaluation" – is a decision that combines scientific, economic, public and regulatory considerations.

All of the elements of a risk evaluation were present in the pentachlorophenol case. For example, expert committees were used by the various government departments. Scientists engaged in a debate in the academic literature. A scientific assessment of that literature took place in several regulatory settings. Public hearings were held in several countries. The economic aspects of pentachlorophenol use were taken into account. As one federal official in Canada said:

> Unions don't realize that there are no suitable alternatives that we would consider for registration. It will have to be a compromise. It is a case of putting enough pressure on to find alternatives. (There is) an unfortunate data gap. Alternative companies have to get the data base to get their products registered. The market ramifications are great, but they (companies that might produce alternatives to pentachlorophenol) aren't moving.

A public advocate group, in the British Columbia case, a labour union, played an active role in the evaluation.

The resemblance between the process of risk evaluation described in the literature, and what happened in the case of pentachlorophenol is a superficial one. The three accounts provided here cannot be integrated into a single picture with any degree of accuracy. From the perspective of those involved in British Columbia, the scientific debate about the Swedish studies happened at a distance, and was not very important. From the Canadian government perspective, the reaction to the Swedish scientists' research in other regulatory tribunals was interesting, but not crucial to their own assessments. From the perspective of the Swedish scientists and of the regulatory tribunals that considered their work, the merit of the Swedish research was the central issue to be determined.

Although they were discussing the same chemical, the subject of scientific debate was also different in the academic journals, in the regulatory tribunals and in the two Canadian assessments. In Canada among regulators and government officials, the scientific debate was about the methods for the assessment of dioxins and the impact of dioxins on health and the environment. In British Columbia, the debate was about the merits of different methodologies used in epidemiological research, and the exposure of the workers. In the regulatory tribunals outside Canada, the scientific question was whether to take the conclusions of the Swedish studies seriously. In the academic literature, the debate centred on the relative merits of different techniques used to control for bias in epidemiological research. Obviously, all these assessments touched upon similar issues, but the focus of their attention was different.

No public hearings were held in conjunction with the pentachlorophenol controversy in British Columbia. Would the situation have been changed with public hearings about the scientific assessment or about regulatory options? Probably not. The kind of information that was being taken into account by the union was not likely to be discussed in public hearings about the regulatory status of pentachlorophenol. This information concerned the actual working conditions in the mills, the state of inspection and regulation in British Columbia, the "good will" or lack of it on the part of industry, and the status of its own study in the conflict. Even if the hearings had dealt with general matters concerning regulation, it is unlikely that these issues – which are central to the union's involvement – would have been addressed.

The presentation of the case of pentachlorophenol as three separate accounts underscores the difficulty of arriving at a rational and sequential process of risk assessment. As a consequence, it raises questions that will be addressed later about how to develop standards. It also suggests that caution is required in dealing with the relationship among science, values and public policy that is characteristic of mandated science. Easy solutions to the dilemmas of mandated science are not to be found in the design of institutions that integrate scientific and regulatory evaluations into a rationally segmented and sequential process.

*Scientists in a regulatory debate*

The Swedish scientists express bewilderment at the reception of their work in the regulatory arena. Their perception is that the regulatory review process was unfair and unscientific. On the basis of examining the correspondence and testimony, we think their perceptions are justified. Even if the Australian case is considered an aberration, the reaction of the regulatory community to the Swedish research should be explained.[28]

To begin, it should be stated that the conclusions of the initial studies, their later analysis and the existence of contradictory findings are not particularly noteworthy in a scientific context. Problems of how to gauge exposure to a chemical hazard confound all epidemiological research, especially when a long latency period is involved. A variety of techniques – each with methodological problems – have been adopted to deal with these problems. We stress that the approach taken by the Swedish studies was neither unusual nor unorthodox.

There are several factors that led to the negative – in some cases, hostile – reception of the Swedish studies in a regulatory context. First, the Swedish studies were seized upon by citizens' groups, including the union in British Columbia, because they provided a relatively unambiguous foundation for the fears about pentachlorophenol. The fact that the Swedish scientists were outsiders to the regulatory process meant they were to be trusted, since regulators were not. Once the Swedish scientists were "adopted" by the advocate groups, their own independence was questioned. The studies became protagonists in the conflict, regardless of the intentions of their authors, and this affected regulatory perceptions of the research. The same phenomenon was evident in the Toronto lead controversy, with respect to the research by the university scientists.

Second, although the Swedish scientists thought of themselves as conventional scientists, and as abiding by the norms of conventional scientific work, they became entangled in the highly polarized regulatory conflict about pentachlorophenol, and about 2,4,5-T and Agent Orange. Increasingly, in response to criticisms of their work, their comments became political, especially as the comments were understood in the emotionally charged atmosphere of regulatory assessment.

For example, the Swedish scientists identified common passages in the submission of the Monsanto company and the final report of the Australian Commission. This was interpreted as political. When the Swedish scientists responded to the accusation that their relationships with the media biased their study, they stated that they had a *moral* obligation to speak with the press. Again, this was seen as a political comment. Most of the Swedish scientists were newcomers to the regulatory debate, inexperienced in gauging how statements like these might be understood. An open association of morality and independent science undermined their claim to be dispassionate scientists.

Third, some aspects of the review process contributed to the hostile reaction to the Swedish studies. How much did the JMPR draw upon others' assessments in its evaluation of 2,4,5-T? Probably a great deal. Did the European Common Market expert committee draw upon the regulatory assessments conducted in its member countries, or even in the United States? Probably. The Director of the *International Register of Potentially Toxic Chemicals* was quite frank. The *Register* did not constitute an endorsement or a separate assessment of the chemical involved. Yet, how many people regard the *Register* as an independent source of information about toxic chemicals? We suspect many do.

The Kilpatrick report found its way across the Atlantic, and it was relied upon in a civil trial launched by citizens groups about the herbicide 2,4,5-T. The scientists who participated in the Australian Royal Commission were mainly the same scientists who had appeared elsewhere in regulatory evaluations of these chemicals. Reliance upon the same network of expertise in a variety of different inquiries and by different regulators created the impression of many different assessments. In reality, a single review of the Swedish scientists' work was being conducted at different times and places. Their research program evolved during this period, but the assessments of their original studies did not.

Fourth, the Swedish scientists were epidemiologists, while most research evaluated for the purposes of regulation is toxicological. Epidemiologists are quite frank about the limitations of their research. It is widely recognized that a variety of innovative techniques will be required in order to obtain reasonably accurate measurements of exposure and harm, and that considerable judgement is required about which techniques to use. The methodological problems of the Swedish studies were particularly noticeable to assessors who were mainly toxicologists, or regulators used to dealing primarily with toxicological data. Although toxicology is beset with its own methodological problems, those of epidemiology seem particularly noteworthy to assessors used to dealing with the results of controlled experiments in laboratories.

Epidemiological work appears to gain acceptance, especially in regulatory circles, only when groups of related studies converge to similar conclusions. It is a maxim in the field that any single epidemiological study will be subjected to severe criticism. One need only to reflect upon the long history of attempts to demonstrate a link between cigarette smoking and lung cancer. It was only after a multitude of studies that the link was established to the satisfaction of the regulators, and even of scientists and the medical profession. Although the Swedish scientists may never be vindicated, the vehemence of the controversy in the regulatory arena and elsewhere is not particularly surprising.

At the same time, epidemiological research is both intelligible and appealing. The results are taken from studies of human populations, not animals. The measurement of exposure is directly related to the actual conditions in which the exposure occurs. The ailing workers are usually there to be counted; their symptoms are obvious. Relatively little extrapolation is required in applying the results of epidemiological research to public policy.

Regulators have another reason to be wary of epidemiological studies as a consequence. Their publication seems to demand immediate action, but other constraints on the regulatory process usually preclude it. The intelligibility and appeal of epidemiology to the public is likely to ignite public controversy, and scrutiny of their own regulatory actions. The direct application of its conclusions to public policy leaves little room for scientific debate, or at least, little public tolerance for further study and delay.

In the case of the Swedish studies, the situation was complicated by the association of pentachlorophenol with Agent Orange and the existence of several outstanding legal actions. A decision favouring the conclusions of the Swedish studies was likely to be costly for other groups beyond the forest industries who use pentachlorophenol. If the attitude of the regulatory tribunals to the Swedish research seems unduly skeptical, given the academic credibility of the research, it is not surprising.

### *Science and the burden of proof*

We began this chapter by saying that no one questioned the danger posed by pentachlorophenol. We end it with less certainty than we began. Accepting the Swedish studies on their authors' assessment of their significance, we still have more questions now. We are not industry scientists, whose mandate is to deal with dangerous aspects of chemicals within a context that supports their continued use. We are not government scientists working in a department with split commitments to regulation and economic development. We have not sat, as members of agency tribunals do, listening to industry argue about the importance of pentachlorophenol. We are simply social scientists with more than the layperson's familiarity with the scientific literature.

We have alluded to the situation we now face at several points in this chapter. We have said that conventional scientists function with a deeply ingrained capacity to handle ambiguity and methodological indeterminacy. Were we functioning in the world of conventional science, we think that our current questions about the hazardous nature of pentachlorophenol would be of no particular significance. We are not functioning in such a world. By studying pentachlorophenol – and the controversies surrounding it – for more than two years, we, and others, have come to expect that we should "take a position" with respect to the toxicity of it. We, too, then, have become like mandated scientists.

In doing so, we bring quite different criteria to the evaluation. We began with a simple answer – and a reasonably defensible one – and end up with a number of questions. We begin by knowing that pentachlorophenol is dangerous, but conclude with an appreciation of what remains to be learned about its toxicity. We end with "reasonable doubt", much in the same manner we would if we had been listening to lawyers and expert evidence in a courtroom. We have come to think of the evidence about pentachlorophenol (and about lead pollution) as if it were evidence in a trial about murder. We have heard the equivalent of how the blood on the floor could have come from many sources, how the smudged fingerprints cannot be identified without question, how the defendant was seen at another location, how his friends do not believe he is capable of the deed.

This is not a courtroom however, and neither is the situation of mandated science. For in mandated science, there really is no jury or judge empowered to weigh the evidence and make the final decision according to some long established rules about how to arrive at such judgements. If we ourselves are like a jury member, we have no judge to lecture us on which evidence should be disregarded, and no collective process by which we will retreat, consider the evidence and make a single judgement. Unlike in the courtroom but in a similar manner to mandated scientists and regulators, our decisions are not final, but one more round in a continuing negotiation – and court cases, in the instance of pentachlorophenol – among industry, public groups and regulators about the regulatory status of a chemical.

We are no longer functioning exclusively in a scientific setting either. Questions that might be left unresolved for the time being, demand resolution so that regulatory actions can proceed. Research is perceived by participants in the controversy as a delaying tactic, and these perceptions are reasonable from their perspective. The complexity of the issues revealed by additional research disguises a simple conclusion: pentachlorophenol is not safe when it is used improperly. There is a great deal of evidence (of highly unsystematic nature) that pentachlorophenol is not yet being used with due caution. Ironically, this type of evidence seems to have no place within a regulatory assessment of a chemical hazard, but it is often uppermost in the minds of the groups who initiate controversy over the regulatory status of a chemical.

Chapter Seven

# Standards Revisited

Each of the four case studies illustrates different aspects of standard setting. In the last two chapters of the book, the material will be brought together as an analysis of mandated science. In this chapter, the emphasis will be on standards and standard setting.

The analysis will begin by returning to our original definition of standards, and providing a more extensive discussion of their characteristics. We will compare the work of the two standards organizations and the role of standards in the two controversies that we have studied. This will permit us to develop the arguments introduced in the chapter one about the relationship between standards and economics, standards and the legal process, and standards and values. We will then return to the debates about standards that were introduced in chapter two, re-evaluating their significance for mandated science. Finally, in this chapter, we will examine the relationship between standards and mandated science, as an introduction to the final chapter on mandated science.

## The Characteristics of Standards

In chapter two, we noted that standards have been developed for almost all products that are exported or imported. They exist for most industrial processes, and are used by engineers and architects as well as industrial hygienists. We have suggested that standards are simply norms governing economic activity, and that the norms are applied to products and activities that exhibit a range of differences. They are designed to identify an acceptable range of deviation in the level of performance. The measurements used in conjunction with standards are usually averages. For example, the level of airborne contaminants in the factory will fluctuate; the amount of pesticide residues on apples will reflect the rate of their deterioration in storage or transit. Yet, both might easily conform to the standard.

We noted that the norms used as standards differentiate acceptability from unacceptability. Acceptability is something other than either excellence or desirability. A kettle that conforms to the standard might be unexceptional in its design or performance, and it might not be particularly desirable as a kitchen appliance. Neither the air in the factory nor the apple sprayed with pesticide necessarily reflects the best quality that can be achieved, yet both might meet the prevailing standard. The term acceptability refers to specific decisions that are made by different groups, and not to a general or societal notion of what is an acceptable level of risk.

Our research suggests that acceptability usually does not reflect public perceptions of acceptable risk. This observation runs contrary to some commonplace views about standard setting, and about the importance of public perceptions of risk. Our argument is as follows. Relatively few standards, world-wide or in any particular jurisdiction, are set with a full public hearing process. Most standards are also not very controversial. Public perceptions are only articulated and represented forcefully when there are opportunities for participation and when chemicals are particularly controversial. The inclination of and opportunities for the public to express its views are limited, even when public hearings are used.

Given these limited opportunities, it would not be surprising if extensive public involvement occurred only in the case of a few standards. For standards that are neither controversial nor debated in public hearings, the role of the public perception of risk is minimal, and is limited to the extent that government officials generally take public opinion into account. Our case studies suggest that public perceptions of risk were not very important in the assessment of specific chemicals, although they did serve as catalysts for the review of particular standards, particularly of standards that were in use for some time.

In our case studies, decisions about acceptability were made by industry, by standard setting organizations and by governments. Acceptability was different in each case. In the case of ACGIH standards, acceptability reflected the decision of the TLV committee, which weighed several different factors, including the field experience of the industrial hygienists, before reaching its recommendation. In CCPR, acceptability reflected an assessment of both health and trade issues, the dual mandates of Codex. Other organizations, such as the European Common Market, arrived at different views of acceptable, reflecting their particular priorities. Thus the mandate of the standard setting organizations is important because it provides their criteria for determining acceptability.

Finally, our research indicates that there are many different kinds of voluntary standard setting organizations and many different kinds of regulatory standards and guidelines. Some standard setting organizations relied upon experts, and others relied upon a consensus of the interested parties to develop their standards. Industry standards differed from those of ACGIH. Except where they had been adopted from ACGIH, OSHA standards were different from those of ACGIH. Canadian and American authorities relied upon a similar body of scientific information to develop standards, yet their standards often differed.

Sometimes these differences reflected scientific judgements about how to interpret the research. More often, they reflected the fact that different criteria had been brought to bear upon the assessment, and thus, the mandate for each group.

The existence of different standards from different organizations – or in different jurisdictions – is important. It creates a lobby for increasing or reducing their stringency. On the one hand, the companies used the existence of more lenient standards in other jurisdictions as an argument for reducing the stringency of specific standards. They also argued that stringent national standards prejudiced their trade relationships with countries that do not abide by the same rules. In this sphere of activity, as in many others, they argued for a "level playing field". On the other hand, advocate groups seeking more stringent chemical control often pointed to examples of countries that have banned particular chemicals as proof that more stringency was needed.

## The Character of Standard Setting: the Two Organizations

In spite of their obvious differences, a great deal can be learned from CCPR and ACGIH about the common characteristics of standard setting organizations. First, both organizations considered their work to be public, but both were public in a specific sense. Neither used public hearings; both relied upon expert evaluations rather than public debate to arrive at their standards. Their meetings were open to the public and their documentation was available to the public, but public participation was rare. Neither labour, nor consumer, nor environmental, nor any other advocate group played a significant role in either ACGIH or CCPR. These organizations were public in the sense that they published extensive records of their procedings, including documentation to support their decisions on standards, and that the public was not explicitly excluded from their deliberations. They were also public in that they involve public officials – in ACGIH, only in a voluntary capacity – in their work.

In the debate about standard setting, there are two ways of understanding the idea of a public organization. A limited definition of public is common in many countries where public hearings are rare occurrences. In these countries, a procedure or organization is public if it is operated on a not-for-profit basis, often with the involvement of state officials, and if the documentation relating to decisions is released to the public. By this definition, both ACGIH and Codex are legitimately considered to be public organizations. Public means something else in other countries. In the United States, for example, a procedure is public only if it includes laypeople in the assessment process, and only if principles of due process are applied. From the second perspective, lack of attention to a legally-oriented understanding of due process renders organizations such as ACGIH and Codex undemocratic and less-than-public organizations. In evaluating the work of these two organizations, it is important to know which definition of public is being used.

Second, both organizations maintained a low profile. Many who used their standards have limited knowledge about the standards and the process by which they were developed. This was more true of ACGIH than CCPR, of course, since national delegations participated in CCPR. The two organizations defended their low profile in the same manner. Both claimed that a low profile is required if they were to develop a comprehensive list of standards. Their officials pointed to the limitations of the regulatory process with respect to originating standards.

Third, in both organizations, the voluntary status of their standards disguised their significance. Their standards were guidelines. Neither organization had the power to implement standards or to compel acceptance and enforcement of their standards. Yet, their standards were widely used, and often adopted as regulatory standards. When ACGIH or Codex standards were accepted as regulatory standards, it was often without acknowledgement of their origin. Even as voluntary standards or guidelines, their standards had more legal status than their originators claimed. Once adopted as regulations, ACGIH and Codex standards were often reviewed by the courts. As well, companies cited their participation in ACGIH and in voluntary standard setting more generally as evidence of due diligence in court cases when they were charged with causing harm to health, safety or the environment. Finally, insurance companies used their standards in deciding whether to award compensation, and the appropriate level of compensation, and these decisions were also often contested in courts of law. As a consequence, the voluntary standards of ACGIH and Codex standards had an important legal dimension.

Fourth, these two organizations incorporated science and scientists in setting standards in a similar manner. Both relied upon expert committees, whose members were appointed as individuals and on the basis of their expertise. In both organizations, the expert committees were expected to insulate themselves from regulatory or political pressures. The experts were expected to control their own bias and to put aside regulatory and political considerations by associating themselves with the scientific community and its values. The expert committees recommended the standards. The final decisions about standards were made by organizations that openly combined scientific and non-scientific assessment. In other words, the organization of ACGIH and CCPR reflected the contention that scientific and policy matters can be separated and considered sequentially. We will return to this contention later, and assess whether this separation was actually accomplished in the work of ACGIH and CCPR.

Fifth, both organizations relied upon operationalizations of their procedures that were expressed as the standards. The ADI and the TLV were regulatory constructions. Although they referred to a series of judgements that combined scientific and other types of evaluation, their proper use implied knowledge about the procedures by which they were developed. Without this knowledge, it was very easy to misconstrue the significance of terms such as "acceptable daily intake" or "threshold level". Operationalization is a normal practice in conventional science, but it reflected a specific regulatory imperative in the context of these two standard setting organizations. The operationalizations in JMPR and ACGIH reflected the fact that decisions were to be made with the information available, and under constraints that would have been intolerable in a scientific context. The pressure to work with operational definitions of the key terms was derived from the mandate to develop standards, rather than from the needs of science.

Sixth, in both Codex and ACGIH, industry played a central role. Industry provided the data, and was often involved – directly or otherwise – in the preparation of draft documentation for the expert committees. Industry members sat as observers or consultants on many committees in both organizations. Industry was involved in the determination of priorities for the assessment of specific chemicals. Industry was consulted before standards were adopted. The primary influence was exerted by industry in controlling the data for assessment. Both ACGIH and Codex depended upon the voluntary and willing participation of industry in order to have data to evaluate. Because these two standards organizations were not regulatory, they had no power to compel the release of data. They were very much dependent upon industry for the work that they did.

These standards organizations did not operate under instructions from industry, however. Their own members were not without influence and autonomy. They were often regulators, so they dealt with industry in another context, where a co-operatitive attitude from industry could be rewarded with a sympathetic response to its problems. The regulators in CCPR or on the ACGIH TLV committee also knew when the data were incomplete, for they often had the power to compel its complete submission elsewhere. Industry supported the relative independence of these two standard setting organizations. The fact that Codex standards were often – not always – less stringent than many national ones was enough to make participation in Codex attractive. Industry preferred a single international review, such as is conducted by CCPR, over many national evaluations. Industry had the upper hand, however, in setting the timing and agenda for chemical assessment. By withholding studies or scheduling their release to coincide with a desired priority for chemical evaluation, industry could direct the ebb and flow of chemical evaluation and thus, indirectly, the production of standards.

Finally, in spite of the extensive industry influence in Codex and ACGIH, their members conceived of themselves as "watchdogs" of industry, and in terms of their health and safety mandate. These perceptions were not altogether inaccurate. Standard setting involves a finely balanced equation between health and safety, on one hand, and questions of trade and economics, on the other. The standard setting organizations are the site of the balancing act, but their expert and government participants are particularly conscious of the need for control of potentially dangerous chemicals.

In summary, the two case studies of standard setting organizations illustrate a number of the dilemmas of standard setting. In these organizations, standard setting was a public activity, but not so public as to involve extensive public participation or public hearings. The participants conceived of themselves as "watchdogs" over industry conduct, but they were highly dependent upon industry co-operation. They relied upon experts and the neutrality of science to support their decisions, but they did not permit expert committees to make the final decisions about standards. Indeed, their scientific deliberations were infused with regulatory judgements. Their procedures were easily understandable, yet their use of technical language precluded informed participation by "outsiders". Finally, their standards were only guidelines, but these standards had considerable influence for regulatory purposes and in their own right.

## The Character of Standard Setting: The Two Controversies

The next two case studies are of controversies in which chemical standards played a role. Again, there are many similarities in the two cases. First, in both the Toronto lead controversy and the pentachlorophenol debate in British Columbia, the origin of the standards was obscure and not well understood by their participants. The original numbers were important because they were often adopted or constituted the starting point for a regulatory assessment, and because they were so rarely revised.

Second, in both of these controversies, the actual numbers used as standards were less important than their relative stringency. The importance of this observation should be stressed. Obviously, it matters whether the amount of lead in the blood that is considered acceptable is 30 mg/ml or 300 mg/ml. This is not the point. The information required to determine whether the standard should be set exactly at 30 mg/ml was simply not available. When scientific data were available, they indicated only the approximate range of acceptable numbers with respect to an approximation of harm resulting from exposure in a number of different situations. When and if a standard was to be revised, it was the original numbers that were altered. The decision to be made in these two chemical controversies was about the position of the new standard in relation to the old one.

As a result, in both controversies, the debate was about standards, but it included many issues that are not related to them. The debate centred on the conditions in which the chemical was used, the degree to which the regulators could be trusted, the enforcement of existing standards and the liability of the companies for any damage that was caused by chemical hazards. Scientific data in standard setting were used to determine the desirability of government action. Attention was directed away from the standard itself and toward the manner in which government handled its regulatory responsibilities. Political constraints upon government action, the level of public trust in government regulators and the responsiveness of regulation to the public were all relevant to evaluations of the adequacy of the relative stringency of standards.

Third, the process of revising standards was cumbersome and slow. The standards were not revised simply because new analytical techniques and new scientific studies were available. In the Toronto lead controversy, it took two citizen groups, several court cases, and the vigorous participation of several levels of government to bring about revisions of the standards. In the pentachlorophenol case, a local union acted as a catalyst for change, as did the creation of a new government department in Canada and the controversy about dioxins and Agent Orange in other countries. Even with these pressures, the Workers' Compensation Board standard for pentachlorophenol had not yet been revised at the time this book was written. If these two case studies are typical, public controversy is the most important catalyst in the revision of standards, but even controversy will not necessarily provoke a revision of standards.

Fourth, the similarity in the standards developed by the different organizations was neither coincidental nor a product of scientific assessment. Decisions made by one standard setting organization were used by others. Sometimes standards were simply adopted from one organization by another. This was the case when the British Columbia Workers' Compensation Board adopted the TLVs and changed their name, or when OSHA adopted ACGIH standards. Other times, decisions about standards in one organization acted as points of reference for others. For example, in a TLV meeting, committee members compared their proposed standard with decisions taken elsewhere. An official of an international trade organization described the relationship between standard setting organizations in the following manner:

> You see I happen to be one of those who believe that the major part of the result from international work is not at any time necessarily put on paper.... I think that most of the things that happen would never be put on record. It's just an osmosis of information, if I may call it that. It is how information and thoughts are being conveyed from one individual to another, one country to another...

Fifth, in both controversies, neither science nor scientists were able to resolve most scientific issues. In each case, more scientists and further research failed to provide answers to the simplest questions. Each new study appeared to complicate the scientific issues further. Each new study muddied the waters. Further research drew attention to the areas of inconclusiveness, uncertainty and disagreement within the scientific community about the issues being evaluated. Even if the scientists were not involved in an advocate debate, and if they were capable of maintaining some degree of neutrality in the emotionally charged atmosphere of these controversies, they could not reach a consensus on basic scientific issues. We will return to this point in the next chapter in discussing scientific uncertainty.

In light of this problem, it is important to examine the role played by expert committees, inquiries and task forces in the two controversies. The inquiries and expert committee reports were disappointing. They too seemed to muddy the waters and draw attention to areas of uncertainty. The problems were most apparent in the case of the Toronto lead controversy. It was the intention of the environmental assessment board to consider the scientific issues in the lead controversy in a rigorous and credible scientific manner. In addition to increasing the complexity of the scientific debate, their willingness to allow cross-examination, and the vigorous participation of the lawyers, – especially the company lawyers – meant that the unstated, but real agenda of the hearings was the potential liability of the companies. We contend that a scientific debate did not take place in the environmental assessment board hearings, even though scientists were involved. We shall return to this point later.

Finally, for similar reasons, it will be necessary to return to the question of whether a relatively dispassionate scientific assessment of chemical hazards was possible in any of the controversies we studied. In both chemical controversies, the university or academically-oriented scientists fared poorly. Their research was attacked as biased and lacking in credibility. Their own uncertainties were parlayed into doubts about the worthiness of their research. They themselves were drawn, more or less willingly, into a battle in which science was seen to be an instrument of advocate politics. They were adopted, again more or less willingly, by advocate groups, and their own credibility as scientists suffered as a result.

To deal with all of these observations, we introduced the term "political chemical". A political chemical is one that becomes a lodestone for a broad debate about chemical hazards and their control. Once a chemical became a political chemical, it was difficult to conduct a specific assessment of it. The debates took the form of "set pieces" in a dramatic presentation. The scientific issues, the arguments, their style of presentation, and even the use of experts were similar from one chemical assessment to the next. The specific attributes of a particular chemical became less important than the general debate about chemical hazards.

Several factors were identified that create "political chemicals". One was simply serendipity; another was the coincidence of several groups taking note of different problems at the same time. Association with a major lawsuit or national issue, such as was the case in the connection between pentachlorophenol and Agent Orange, transformed a chemical into a political chemical, even if the issues were different. Frustration with the lack of opportunities for public debate resulted in particular chemical assessments being used for the purpose. Jurisdictional disputes fostered the emergence of political chemicals, as did controversy itself. Finally, the relative intelligibility of the scientific issues connected with some chemical assessments facilitated the involvement of many groups.

## Issues in Standard Setting

In chapter two, we referred to three important dimensions of standard setting as an example of mandated science. In the origination of standards, the relationship between scientific and economic decisions were highlighted. The legal substratum of decisions about standards was relatively easy to identify, and the relationship between science and values was made explicit by the design of risk assessment procedures that attempted to separate the assessment of the scientific and the value-based aspects of risk. In this chapter, we explore the implications of these relationships more fully on the basis of our case studies.

### *Standards and Economics*

It seems self-evident that standards and economics are closely entwined. The cost of installing pollution control equipment, and the problems introduced by standards in trade relations underscore the association of standards with economics. We encountered significant resistance when we asked questions about the relationship between standards and economics. In a few cases, it was vigorously denied that the economic impact of standards was relevant to decisions about standards. A delegate to Codex – an organization dedicated to trade – dismissed the connection between standards and economics in the following statement:

> There is no discussion of economic impact. There is a discussion of science as to whether a suitable limit can be measured, where are the chemical residues, whether it is too high, whether it isn't needed. They (the CCPR) are not discussing the economic impact. It is a scientific discussion.

This statement is disconcerting, because this person is from a country where any regulatory standard – including any Codex standard accepted as a national standard – is subject to a socio-economic impact assessment. Some ACGIH members said the same thing. When asked specifically, some members said that economic issues played no role in the determination of standards.

There seems to be an almost schizophrenic attitude towards economic issues among the participants in standard setting organizations. On one hand, almost every participant attested to the necessity of making decisions that were economically sound. On the other hand, these same participants sometimes denied that economic considerations were taken into account in the development of standards. Why is the link between economics and standards – which appears to be self-evident – so problematic, when the relationship between standards and trade is discussed so openly?

One possible reason for denying the importance of economics in standard setting is simply the scientific orientation of the standard setting organizations. Any reference to economics would corrupt the rationale for standard setting:

"Standards are developed for health and safety reasons and they are based on a scientific assessment." If this were the case, members of CCPR and ACGIH were involved in a conscious practice of deception. They were aware of the economic consequences of their decisions. They openly acknowledged the important role in standard setting played by industry. On the basis of our research interviews, we have rejected the hypothesis that deception – even self-deception – was involved.

A second, more realistic possibility is that the delegates to CCPR were engaged in an ideological battle. The ideology of science served as an instrument for activities that had little connection to science. On the basis of our interviews, we would also dispute this interpretation. It discounts the effort and resources allocated to scientific assessment in standard setting organizations, and the integrity of the experts in evaluating the toxicological data. It also discounts the high degree of sensitivity to the problems of standard setting, and in particular, to the difficulties of controlling industry influence that was evident on the part of many participants.

There is another, more plausible answer to why participants in standard setting might, on some occasions, refuse to acknowledge the obvious relationship between standards and economics. The answer lies in an argument about the nature of the relationship itself. Our hypothesis is that there are aspects of the relationship that cannot be discussed easily. Either they contradict commonplace assertions about standards, or they draw attention to another side of standard setting, one that is not easily acknowledged, or understood.

Our argument rests upon three observations. First, decisions about standards posed some contradictions for industry and government. They privileged some sectors of the economy and some companies, but they also required the active co-operation of all – or almost all – segments of industry. Second, standards operated in restraint of trade. Third, the network of standards and standard setting organizations provided an ill-defined and unacknowledged means of co-ordinating economic relations, a method that was nonetheless compatible with the anarchistic nature of capitalism as a system of economic relations and ideas. These aspects of standard setting provided a powerful incentive for participants sometimes to deny the importance of the relationship between standards and economics. Each of these observations requires further discussion.

The first – namely, that standards privileged some sectors of industry, and some industries – is easiest to demonstrate. If conforming to standards requires significant capital investment, as industries themselves argued, then only some corporations will be in a position to implement standards. It is commonly accepted that larger companies are most likely to have capital resources to implement standards for pollution control, and that standards benefit industries already involved in production. For example, we were told the following story about the standards for vinyl chloride:

> Vinyl chloride is a marvelous example of this. Industry cried and said that it would go bankrupt and disappear if the standard was made as low as it actually was. They have made more money since the standard was lowered. ...Most larger firms have the funds to purchase technology, which is, in many cases, available, to enclose operations with dangerous substances. This is happening more and more, especially in the chemical industry. The situation is different at the level of the small workplace, at the end-user. And these are not places that are of great economic importance to a nation or to a State or to a large political area. You could knock out a few small industries and it would not be very significant.

Companies also use standards (and the proprietary data on which standards are based) as a means of precluding competition. This, too, was commonly acknowledged. Finally, in our discussion of the CCPR, we used several hypothetical examples to show how standards and trade were connected, and how decisions about standards would have a different effect upon the industries that depend upon export trade and the industries operating primarily in domestic markets. This last point was not discussed openly, but we believe it was understood by participants in Codex. Put simply, standards benefit some countries, sectors of industry, and some firms, more than others.

At the same time, decisions about standards require co-operation from almost all countries or segments of industry – indeed, *all* segments, if consensus standards are involved. There are several reasons why standard setting demands co-operation. To the extent that they are adopted as regulations, the standards are applied universally. Standards are also used to determine insurance rates and the level of compensation in the case of injury. They are used by law courts to indicate when due diligence has been taken about workers' health and safety. Standards would only work in trade relations to the benefit of our hypothesized industrialized countries if they are widely adopted. Thus, it is in each country's and company's interest to make sure that as many as possible are involved in their origination. As well, prudence demands that each voice be heard, if at all possible, for standard setting promotes standardization, in the everyday sense of the word.

Our first observation, then, is that there is a fundamental contradiction in the activity of setting standards. While standards are an instrument of aggressive competition and trade relations, standard setting requires active participation – indeed, willing co-operation – from as many countries or segments of industry as possible. We suggest that this duality in standard setting is not easily acknowledged by its participants.

The second observation why it might be difficult to acknowledge the relationship between standards and economics is that standards function in restraint of trade. It is common knowledge among participants in standard setting that the existence of standards can encourage predatory trade practices, or act as non-tariff barriers to trade. Decisions that are beneficial to the economy and its supposedly free markets are also those that restrain trade and encourage "dumping" and predatory trade practices. To the extent that these practices occur, they too pose a contradiction for nations espousing a free market philosophy.

There are some situations in which significant and open restraint of trade is considered legitimate even in capitalist societies with a strong commitment to free market economies. When the standards concern economic functions that are basic to all industrial production (sizing of goods; purity of chemicals), then the implications of participating in standard setting organizations are probably tolerated most easily. For example, although not all countries endorse ISO standards, neither ISO nor its standards are subject to much controversy. When standards are connected to major export commodities, such as food, or deal with major industries, such as produce industrial contaminants, then participation in standard setting organizations cannot be taken for granted. In our research, neither companies nor countries were willing to discuss the implications of their decisions for trade, and the economic reasons behind the arguments they advanced about the appropriateness of individual standards.

Our third observation concerns the relationship among standard setting organizations. We have already suggested that standards and standard setting organizations constitute a loose network of institutions, although these organizations are independent, and often competitive. We believe that this network contributes to the coherence and co-ordination of capitalism as a system of economic relations. In the case studies, the relationship between standards organizations provided the opportunity for face-to-face discussion among those with competing economic and national interests. It provided the opportunity for individual countries, industry organizations or even individual firms to be informed about new developments affecting their products. It provided such bodies with the opportunity to position themselves in relation to others, by the very nature of the decisions made about standards. Finally, the variability in standards and standard setting organizations created a network of choices for individual firms that enabled them to negotiate their relationships within industry and with different governments on terms most favourable to their operations. We have stated elsewhere that "dumping" is no more, or less than the result of choices made about the location of manufacturing operations or the sale of products in relation to the existence of different standards in different jurisdictions.

We think that the term "web" is appropriate to describe the result of all these choices, and of all this standard setting activity. The existence of considerable variation in standard setting provides a structure of choices for the individual firm, albeit different ones in the case of different chemicals. At the same time, standard setting provides a connecting tissue for industrial development. It is a method of co-ordinating trade relations, and it extends, in a manner reminiscent of a web, throughout capitalist societies. The many possible linkages among companies made possible through standard setting organizations contribute to the flow of market relations within and between countries.

If we are correct, the type of economic co-ordination made possible through standards activity does not conform to the usual perceptions of corporate control and co-ordination. Co-ordination occurs through the variety in standards and opportunities for choice that standard setting makes possible. At the same time, in the case of standards, co-ordination is exerted by the activities of many ill-co-ordinated bodies. If standard setting is a co-ordinating mechanism for capitalist relations, it is a particular kind of mechanism. Participation in standard setting for the purpose of co-ordinating trade or economic relations requires little overt intentionality or forethought, and few formal commitments. It is a form of co-ordination and risk management that is compatible with a philosophical commitment to free markets and deregulation. As such, we would argue that it is entirely compatible with capitalism both as a system of economic relations and a system of ideas, for it is anarchistic in practice and in theory.

If these three observations – that standard setting promotes competition but requires co-operation, that standards function in restraint of trade, and that standard setting constitutes a loosely organized mechanism for the co-ordination of relations between firms, sectors and countries within a capitalist economy – are accurate, then it is to be expected that the relationship between standards and economics is a difficult one. It would not be surprising if it were neither acknowledged nor easily discussed by standard setting organizations. To acknowledge the role of economics is to take a much broader perspective than most participants in standard setting organizations would be comfortable with, assuming they would agree with the conclusions we have reached. To discuss the relationship as we have done is to call attention to the contradictions within capitalist economies, and in the way standards are used.

### *Standards and the legal process*

In discussing the common characteristics of ACGIH and Codex, we called attention to two perspectives on the public nature of standard setting organizations. From one perspective, these organizations are deservedly called public. From another perspective, they are not. From the second perspective, public or democratic values appropriate to standard setting should be derived from the legal notion of due process. Standard setting is public or democratic, it is argued, only to the extent that the full panoply of legal rights exists in conjunction with the debate about standards. This requires not only public hearings, but also legal rights for interested parties with respect to their participation, full disclosure of information and cross-examination. From this second perspective, the legal system is used as the point of reference to determine whether a properly democratic process of standard setting has occurred.

The existence of these two different perspectives on the public nature of standard setting is useful, because it permits us to examine the implications of pursuing one of them, the legally-oriented perspective. Before doing so, we should identify our own views. We believe that standard setting can, and should be more, rather than less public. Informed public opinion has much to contribute to the quality of standards, and indeed, without it, many necessary revisions in standards would not occur. That said, the introduction of legally oriented norms, such as due process, to standard setting is not without its drawbacks. These drawbacks as well as other means of achieving a more public process should be explored.

The most useful example from our research of a legally-oriented chemical assessment is the Toronto lead controversy. At least some of the standards in this controversy were guidelines not regulatory standards, and several assessments were conducted without any reference to legal norms, by expert or governmental committees. Yet the environmental assessment board took a legalistic approach to its task, and it permitted cross-examination. Lawyers played a key role throughout the controversy, and in particular in the public hearings. As well, several court cases occurred during the controversy.

We have three observation about the importation of legal norms and values into the Toronto lead controversy. First, reliance upon legal norms did not occur as a result of a search for due process and democratic values. It was primarily instigated by the companies or their lawyers, and it had the effect of curtailing participation from several civic and advocate groups. Second, the debate about standards was, at least implicitly, a debate about whether the companies were negligent and liable for any damage caused by their pollution. This was particularly the case because the companies had already been to court to protect their interests. Even when they appeared before the City Council earlier in the controversy, the lawyers from the companies acted as if they were testifying in a court. They did so because their testimony before the environmental assessment board or City Council could have been introduced as evidence, once the matter was taken to court.

We would suggest that the companies had no choice but to protect themselves by acting as if they were in a court. This point has important ramifications for mandated science. In mandated science, the argument is often made that inquiries and expert committees are preferable to formal tribunals or courts for resolving disputes with scientific elements. In an inquiry, it is argued, the issues can be canvassed thoroughly, and with exclusive reference to scientific norms. Often, for example, the Commissioner of an inquiry will introduce its work by stating that the inquiry has no interest in apportioning blame. The participation of lawyers will be discouraged. The contention is that more scientific discussion will occur if the inquiry rejects court-like procedures.

On the basis of our analysis of the Toronto lead controversy, we would argue that the inquiries and expert committees would have had difficulty avoiding the trappings of a court, and encouraging free-flowing discussion. Only from the layperson's perspective did it seem possible to speak of the issues without focussing on questions of liability and responsibility. Government regulators were obliged to defend their actions and delimit the scope of their responsibility, and industry was obliged to ensure that the so-called free flowing discussion did not create a body of evidence that could have been used later in a less informal, and indeed a legal setting. The legal substructure of an inquiry or an expert committee seemed to be an intrinsic part of its organization.

In a theoretical argument that is useful for our purposes, Thibault and Walker distinguish between truth-seeking and justice-seeking procedures, and they associate the former with science and the latter with the courts and due process. They argue that "a fundamental dichotomy (exists) between the potential dispute resolution objectives of 'truth" and 'justice'". In scientific disputes, they suggest, one is dealing with a "cognitive conflict in a setting of common interest."[1]. Citing Youtz, they then argue:

> Adversary arguments and contentions are only incidental to science, not basic as they are in law. In law, adversary arguments tend to settle differences. In science, adversary arguments send contenders back to investigative procedures with which to prove themselves right – or wrong.[2]

Further to their argument, the norms of science seem to preclude participation of a lay public in scientific assessments. In science, emphasis is placed on peer review as a technique of evaluation, and the criteria for such review are supposed to reflect scientific expertise. Interest group negotiations and adversarial relations are considered inappropriate in addressing the substantive claims of science, whether or not they occur in practice. Truth does not lie in the compromise of interests, although interest groups adjudication may be instrumental in establishing it. Rather, truth can be expressed by someone in a minority interest position.

Legal norms exist for other purposes. Indeed, Thibault and Walker suggest that "the legal process is concerned, for the most part, with conflicts of interest." They note:

> It is true, of course, that the legal process is often concerned with resolving disputes about "facts" and factfinding can be determinant when conflicting justice claims are controlled by established legal rules. But typically, determinations of fact are subordinate to the justice objective. The purpose of factfinding contrasts sharply with the purpose of factfinding in scientific inquiry. In science, facts found have an enduring significance because they guide future conduct.... In contrast, the significance of factual determinations in a legal proceeding generally ends with the division of outcomes and there is no future reliance on the cognitive decision.... The goals are dictated by the underlying character of the dispute and hence cannot be established independently.[3]

The representation of interests is an essential element of the legal process. The fullest possible representation of those interests, in the fairest possible procedure (including due process, representation of all interested parties, full disclosure of information, cross-examination etc.) constitute the goals for legal proceedings.

Thibualt and Walker's analysis is useful because it places emphasis on the objectives in a dispute, and the role of factfinding and procedures in achieving different kinds of objectives. They stress the differences between truth-seeking and justice-seeking proceedings as a way of developing a theory of procedure that will provide guidance in resolving "mixed disputes", those involving both scientific and legal issues. The primary difference in procedure in the two types of disputes lies in the role of interests and interested parties in their proceedings, and of compromise in their resolution. The problem with mixed disputes is that both types of procedures are required but there is a fundamental dichotomy between them.

Our third observation from our research draws upon Thibault and Walker's analysis. The resolution of both "truth" and "justice" disputes depended upon full disclosure of information. We observed, however, that in a scientific dispute where interests were involved, as they were in the Toronto lead controversy, it was unlikely that participating scientists would participate in a full disclosure of information without some form of adversarial proceeding. With the partial exception of the university scientists, these scientists were used as expert witnesses. A decision had been made by the interested parties about whether their research would be submitted for assessment, and whether they would be asked to testify. Like all witnesses, the experts were interviewed before they gave testimony. At least one scientist did some work for the companies but was not asked to testify. It is not likely that the expert witnesses would have falsified their data to suit the interests of those who hired them, but circumstances decreed that they were also unikely to speculate broadly on the witness stand about potentially contradictory findings. Thus, only an adversarial proceeding could have elicited all relevant aspects of the scientific information, because only cross-examination could have elicited information or uncertainties that were not forthcoming otherwise.

It is easy to ignore the quite fundamental differences between truth-seeking and justice-seeking procedures, or to think that the components of a mixed dispute can be separated and dealt with sequentially. This is the dilemma of mandated science, which is characterized by a high proportion of mixed disputes. Like other scientific inquiries, mandated science is intended to be truth-seeking in its objectives, orientation and procedures, but the procedures appropriate to the resolution of truth-seeking disputes do not take into account the conditions under which mandated science operates. Abandoning a truth-seeking approach for a justice-seeking one is equally problematic, for a compromise of interests will not necessarily produce a safe working environment or adequate decisions about a pesticide.

Thibault and Walker proposed a sequential assessment of risk as a means of handling the truth and justice objectives in mixed disputes. Their proposal is similar to many others, and indeed one that has been implemented in several countries since their article was written. In formal risk assessment procedures, "truth-seeking" and "justice-seeking" are associated with two different stages of the same assessment procedure. In fact, attempts have always been made to separate the truth and justice seeking aspects of chemical assessments. For example, in ACGIH, the TLV committee is supposed to conduct the scientific assessment, while the organization's annual meeting approves the actual standards. Codex uses three three organizations, CCPR, JMPR, and Codex Alimentarius, to ensure that both truth-oriented and justice-seeking interests (in this case national interests) are satisfied.

We are not convinced that the combination of truth-seeking and justice-seeking procedures in risk assessment is particularly successful in any of these examples. In the next section, we will provide the reasons for our evaluation.

*Standards and social values*

Risk assessment is often described as a procedure for bringing together a scientific and a value-based assessment. The first stage of a risk assessment, whether it be the formal procedures adopted in several countries or the informal arrangements pursued in ACGIH and Codex, is intended to deal with scientific issues and to be conducted according to the canons of the scientific community. The second stage is intended to deal with explicitly political and value-based questions, which are not amenable to scientific dispute resolution. For reasons we will discuss more fully in the next chapter, we prefer to describe the two stages of risk assessment as truth-seeking and justice seeking, rather than as scientific and value-based, and to focus directly on the relationship between the objectives of the assessment and the procedures used to achieve them. It should be noted, however, that in evaluating the resolution of the mixed disputes in standard setting, we are also dealing with some aspects of the relationship between science and values in mandated science.

In our case studies, the sequential assessment of risk was not very successful for a number of reasons. First, in none of the controversies that we studied was the situation condusive to a sequential assessment of risk. In the pentachlorophenol case, we were forced to provide three different chronologies, simply because any attempt to combine them would have distorted the picture from each perspective. We think that the pentachlorophenol controversy is typical of others. Events and decisions in this controversy could not have been forced into a rational pattern, even with the introduction of a formal risk assessment procedure, because of the jurisdictional complexity of chemical assessment, the serendipitous choices of the participants, the different agendas and preoccupations of the various participant groups, and the "spill-over" effect of the controversy itself.

We are equally convinced that the assessments in ACGIH and Codex did not conform to a sequential pattern, although the formal origanizational structure in each case was designed to achieve one. We draw attention to the role of industry in the scientific component of their assessments, both in terms of setting the agenda for assessment and in terms of providing the necessary information. We have already discussed the nature of the operationalizations in the terminology used by the scientific committees of each organization. Terms such as ADI and TLV relied upon by the JMPR and the ACGIH TLV committee in their scientific assessments reflect both scientific and regulatory judgements. We have argued that in spite of their attempt to insulate scientists, regulatory pressures impinged upon the decisions of these expert committees. Thus, we believe that the first stage of the assessment in both ACGIH and Codex cannot be described accurately without reference to its non-scientific aspects.

In standard setting generally, we have observed that decisions usually concerned the relative stringency of a proposed standard or its revision. In dealing with relative stringency, it was necessary to take into account many aspects of the political and regulatory process that were used to evaluate the adequacy of one standard as opposed to another one. We also drew attention to what we called political chemicals, arguing that it was political chemicals and controversy that stimulated the revision of standards. The scientific evaluation of political chemicals is inevitably compromised by the assessment of the regulatory process that accompanies it. Finally, the experience of the academic scientists before the lead and pentachlorophenol tribunals should serve as a caution about the status of the "truth-seeking" stage in risk assessment, for these scientists did not consider their experience to have been a scientific one.

There is another, quite different reason to question the success of a sequential process of risk assessment on the basis of our research. The point is best made by examining the controversy over pentachlorophenol in British Columbia. The labour union in British Columbia saw itself as engaged in a scientific debate about the best method for gauging exposure of its workers to pentachlorophenol and the effects of exposure. On one hand, this involved relying on epidemiolgical research, for such research could indicate whether pentachlorophenol exposure was linked to the symptoms of ill-health experienced by workers.

On the other hand, the union wanted an appreciation of the actual conditions in the mills to be included in the assessment of risk. From their perspective, risk could not be measured unless one took account of how the chemical had been, and was being used. The union argued that the actual risk was greater if a chemical was used improperly, as well as if it was potentially a more potent carcinogen. In other words, the union wanted a site-specific evaluation of risk. This type of evaluation is not easily accommodated in the scientific component of a risk assessment, where reliance is usually placed upon toxicological data. It would not be measured in an epidemiological study either, because epidemiology deals with the conditions of exposure after the fact.

More significantly, it was also a type of assessment that bore directly upon the interests of the workers, and one which required sensitivity to questions of fairness and justice. Conducting such an assessment meant stepping into the vortex of labour-management relations, and evaluating the working conditions and labour process used in the pulp mills in British Columbia. It would have been considerably more difficult to maintain the seeming neutrality of science in such an assessment, however much care was taken to be rigorous in the design of the research. The very conduct of research would have affected the interests of the parties in dispute in the controversy.

Our point in this and previous examples is not to suggest that scientific and value questions can never be distinguished to any reasonable degree. The focus of our attention is on the procedures used to resolve disputes in mandated science that are inherently mixed ones. The problem of mixed disputes is endemic to mandated science. In site-specific risk assessment, in the evaluation of the relative stringency of a particular standard, in the actual practices of ACGIH and Codex and in the controversies we studied, it was very difficult to conceive of how the disputes could be "unmixed". Thus we find the proposals for a science court made by Thibault and Walker and others for a sequential risk assessment process or for such innovations as a science court to be less useful than their analysis of procedures. Our contention is that reliance upon a two-stage approach to risk assessment does not resolve the dilemmas of mandated science, nor does it address the problem of how to handle mixed disputes.

Like Thibault and Walker, our empirical evidence suggests that neither the importation of the norms and values of science into a legal process nor the importation of legal norms into a scientific assessment will resolve mixed disputes. Unlike them, and on the basis of our research, we do not believe that such disputes can be disentangled into their truth-seeking and justice-seeking components. Our conclusion is that standard setting requires its own norms and procedures of evaluation. Considerably more thought is required to develop them than has been displayed in the literature on risk assessment procedures to date.

## Standards and the Debate about Regulation

We referred earlier to two debates about standards in the academic literature and policy arena. The first debate concerns the relationship between standards and regulation, and casts the issues as a choice between deregulation and government intervention. In this debate, the standards being discussed are regulatory standards. Deregulation would involve dismantling many of the government regulatory agencies, and substituting market-related mechanisms in the place of regulatory standards in order to protect the environment and health and safety more efficiently.

In the debate about deregulation and government intervention, the choice has been presented as whether the public or the private sector should be responsible for standards and their implementation. The problem with viewing the choice in this manner – as being between regulation and deregulation – is that it neglects the reality of standards, which are neither fully public nor fully private in orientation. Standard setting always involves some degree of co-ordination between the public and private sectors.

ACGIH, ISO and CSA are thought of as private sector organizations. In fact, they are quasi-public organizations. In the case of CSA, government representatives are one party to the consensus on CSA standards. CSA standards are also adopted by Ontario Hydro and used by it as regulations. In effect, if not in law, CSA and Ontario Hydro operate as a partnership. ISO is a private organization, but some of its members are from government-sponsored standards councils. ACGIH draws its primary membership from governmental industrial hygienists. Even the industry-based standards organizations operate on a non-profit basis, and have public service mandates. Are all these organizations to be considered to be exclusively in the private sector?

On the other hand, regulatory standards are less public than they first appear. Industry plays a significant role in developing them, by virtue of its proprietary control over the data, and its continual presence in the deliberations concerning the standards. Every regulator knows that a good deal of negotiation goes on between industry and government about standards and their enforcement. Currently, interest is being expressed in several countries in environmental mediation, a government supervised consensus procedure for developing standards with the participation of industry and various environmental groups. Again, should this form of standard setting be viewed as exclusively public?

The debate about regulation and deregulation and about the value of regulatory standards is about the subtleties and complexities of the various types of relationships between the public and private sectors with respect to decisions about standards. This is not usually an open debate, for none of the options – whatever their mix of public and private participation – conforms to public expectations, either about a fully public and vigorous process of standard setting, or about the importance of a free market approach.

The proposal to make standard setting exclusively public in orientation is a radical one. It would require government testing of chemicals, and a dismantling of the proprietary nature of the information used in chemical assessment. It would require forced disclosure of information, extensive record keeping and an active, properly funded inspectorate. It would require regulators to be cordoned off from the industries they regulate, a proposal that alters the nature and function of regulation. Perhaps this is preferable to the current situation, but it is wrong-headed to believe that these changes can be accomplished by a modest program of reform. The current reality of standard setting is far from a public one.

The alternative proposal is to leave it to market mechanisms to ensure health and safety are protected. These proposals are equally radical, for they neglect the important aspects of the standard setting process as it now occurs. Industry has been an active proponent of standards, because independently developed standards assist industry in managing its risks, through insurance, in the law courts and in trade relations. Without standards, insurance could not be priced or compensation awarded. The legal defence of due diligence would be much harder for companies to establish. Only seemingly neutral standards serve these purposes adequately. Public or governmental involvement in standard setting is essential in rendering standards useful and attractive to industry.

From the perspective of standard setting, the debate about regulation and deregulation is useful noise in the system. It is useful in that it crystalizes some trenchant moral critiques of government or industry involvement in standard setting. It is noise in the system because the real choices lie elsewhere. The real options concern the combinations of public and private involvement in standard setting. The real debate takes place in much less public forums than the debate about regulation and deregulation, and it is a struggle for different forms of compromise between public and private sector needs.

### The Debate about Standards: Prescriptive versus Performance Standards

The second debate about standards concerns the relative merit of prescriptive versus performance standards. Prescriptive standards are said to be regulatory in orientation, while performance standards are seen to be deregulatory. The debate about the relative value of prescriptive and performance standards does have significant implications for regulation, although not simply in terms of a choice between regulation and deregulation. The problem with the distinction between prescriptive and performance is a familiar one. Just as standards are rarely ever completely public or market-based, so too, standards are almost never clearly prescriptive or performance. Yet the debate about prescriptive versus performance standards is useful to its protagonists.

A genuine performance standard would be derived from biological and genetic screening, for such a standard would connect an exact level of exposure to a specific level of harm from it. But biological and genetic screening standards are yet to be developed, and may never be, because of the significant ethical and civil rights issues connected with them. A true prescriptive standard is more common. We used height restrictions on smokestacks as an illustration of prescriptive standards. Our example was misleading. There are standards for the height of smokestacks, but most of these concern the impact of smokestacks upon the skyline of a city. More often in practice, attention is directed to the emissions from the smokestacks, or the contamination of the air as a result of them. In both emission and contamination standards, some attempt is made to assess the extent of exposure and the potential for harm resulting from it. Both of these assessments are more properly associated with performance standards.

Many of the other so-called performance standards are not true performance standards. For example, biological limit values used by ACGIH are considered to be performance standards, but they do not connect the harm caused to particular individuals to the specific levels of exposure to contaminants. Impingement standards are seen to be performance standards, but this terminology is correct only in the most general sense. Impingement standards do not purport to identify the relationship between exposure to a chemical hazard and the harm caused by it. We have referred to the TLVs as performance standards. The TLVs are concerned with human exposure to chemical contaminants, but in the case of TLVs, only a limited attempt is made to assess the actual human harm resulting from chemical exposure. The TLVs are time-weighted estimates of the level of contaminants in the air, rather than measurements reflecting actual exposure and the harm caused by it.

Debating the relative merit of prescriptive versus performance standards is not very useful, if what is at issue is a choice between two options. At the same time, the debate itself serves two important purposes. It is both a debate about how responsibility for harm will be assigned and a debate about who should bear the risks involved with chemical exposure. In both cases, the question is which specific type of standard should be adopted, and the options reflect decisions about how risk and responsibility for harm will be assigned in each case.

It will be useful to return to the Toronto lead controversy, and the judge's decision in order to see how responsibility is allocated differently with different types of standards. Recall that the judge was dealing with an impingement standard, but one in which the requisite elements of a legally-convincing argument were difficult to provide. In the Toronto lead controversy, the Ministry of Environment could not prove that the emissions caused the elevated blood levels, nor could it be certain that the actions of the companies created the elevated blood lead levels. None of the Ministry's own tests established the link between exposure and harm. Indeed, for every expert witness the advocate groups introduced to establish that harm had occurred as a result of exposure to lead, the companies responded with an expert witness who claimed the opposite.

In contrast, a genuine prescriptive standard would have been much easier to use in the legal proceedings of the Toronto lead controversy. Violation of the standard (the height of the smokestack, the amount of contaminant emitted from the smokestack) would have constituted the *act*. The *actor* would have been the companies involved in the violation. The *act could have been attributed to the actor* without much difficulty. The assessment of *harm* would have occurred before the violation, and would have been based on extrapolations from toxicological data rather than the experience of the citizen residents. The *connection between the harm and the violation* would have been much easier to establish, assuming that an accurate record of the violations had been kept.

The advocate groups did not want to use a prescriptive standard, however. Their argument was that at least some of the contamination came from fugitive emissions, not the smokestacks. They also argued that the companies would simply reduce their output of contaminants from time to time or when measurements were being taken so as to achieve an acceptable *average* level of contamination. Their interest was in the people who had purportedly been injured, and in linking the exposure caused by the companies to the actual harm resulting from it. Given the difficulty of using a performance standard in a legal setting, they sought legislation that would place the burden on the companies to prove that they had not caused the harm.

The debate about the relative merit of prescriptive and performance standards is useful, because it highlights the fact that different types standards are more or less useful in a court of law if responsibility for harm is being determined. To the extent that any standard diverges from the genuine prescriptive standard, it is more difficult to use it for this purpose, unless the legislation places the burden of proof upon the company to show why it is not responsible for harm. The debate about the choice between prescriptive and performance standards is misleading, because it implies that a simple choice between the two alternatives can be made. Each type of standard differs from others in terms of how it facilitates the identification of the elements of a legal proof. The TLVs are usually called performance standards, and Codex standards are prescriptive in orientation, but neither can be used easily in a legal context when responsibility for causing harm is being determined.

The various types of standards also differ with respect to how they identify and allocate risk. Consider the following examples of standards. With the smokestack standard, the genuine prescriptive standard, it is relatively easy to determine who bears the responsibility for the risk when such a standard is violated. In this instance, the company who owns and operates the smokestack bears the legal risk for violating the standard. The smokestack standard can be compared with the impingement standard, where determining responsibility for causing risk is more difficult. As was evident in the Toronto lead controversy, it was very difficult to identify who caused the risk with the use of an impingement standard, since the lead could have come from several sources. With biolological and genetic screening, the purpose of the measurements is to determine whether an individual has a particular predisposition for risk. Whose responsibility is it to ensure that such a risk is avoided – the worker or the potential employer?

In very general terms, the allocation of risk with different standards is similar to the allocation of responsibility. Prescriptive standards locate the risk with the company whose activities are being regulated. Performance standards (especially codes of conduct) are less likely to do so. More often, they locate the risk with the individuals who have assumed it, either by being party to a code of conduct, or by accepting employment in a risky situation. This fact underlies the connection between prescriptive versus performance standards and the debate about regulation, for those supporting deregulation often support the use of codes and of performance rather than prescriptive standards.

## Standards and Mandated Science

Mandated science involves the relationship of science, values and public policy. The study of standard setting suggests that this relationship is usually defined much too narrowly. Standard setting involves science, values, public policy, and legal, trade and economic relationships. We believe that it cannot be understood without paying attention to each of these relationships.

We have observed that even the term "science" was understood in two different ways in standard setting. On one hand, "science" referred to the evaluation of toxicological and other research studies by expert committees and appropriately trained individuals. This "science" was easily recognizable by members of the scientific community, even if the assessment occurred under constraints that they might have found unacceptable. On the other hand, "science" referred to the product of such committees as the JMPR, their recommended standards. This latter "science" was quite different from the activities normally associated with the scientific enterprise, whatever its merits on other grounds. It was a label for a number of techniques that were used by expert committees to deal with inconclusive and conflicting data, with factors that were not scientific and with the combination of scientific and regulatory judgements. These techniques were quite different from the normal practice of science. In evaluating the relationship between science, values and public policy in mandated science, it was obviously important to keep these two "sciences" in mind.

The value debate was also more complex in standard setting than it is usually portrayed in discussions of mandated science. Ethical values, economic and political interests and legal norms involving fairness, due process and natural justice were all involved in mandated science, and the relationship among these different values was complicated. The tensions between scientific and legal norms were not easily resolved, and we believe that they could not be addressed successfully by simply combining them in a risk assessment procedure. Trade and economic relations associated with standards could not be described by simply referring to the interests of participants in standard setting. The existence of political chemicals complicated the value debate about chemical hazards and their control significantly, in terms of the range of values brought to bear in any particular evaluation.

When we began the research on mandated science, we conceived of the relationship between science, values and public policy in fairly simple terms. We might have represented it in the manner illustrated in Figure 1.

At the conclusion of the study of standard setting, we believe that the relationships of mandated science in standard setting, should be illustrated more like Figure 2.

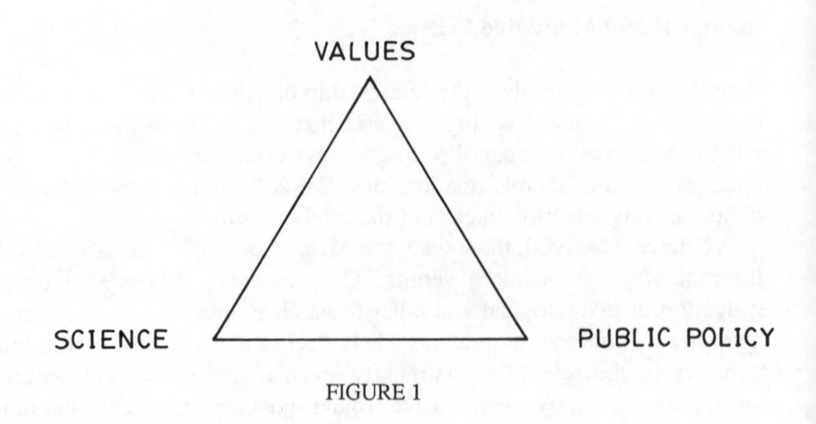


FIGURE 1

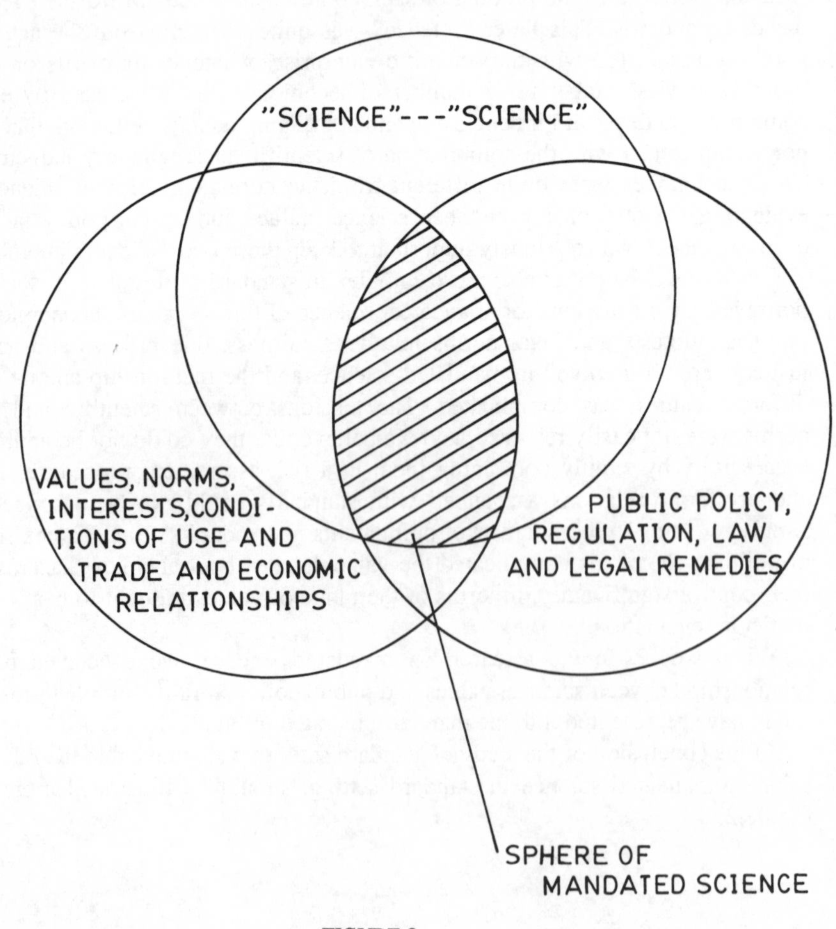


FIGURE 2

A number of proposals for the reform of standard setting have been advanced. We will not discuss them here, partly because they often involve decisions relevant to specific countries. The problem with most of these proposals is that "trade-offs" are involved in standard setting. The process can be made more public, but less comprehensive. The standards themselves can be made more comprehensive and deal with environmental matters more generally, but if standard setting organizations are not also trade organizations dealing with specific commodities, the production of standards is slow and their impact is minimal. Industry exerts an undue influence in the origination of standards, but without industry's active and willing co-operation there would be no data to review, and no standards to criticize. Public hearings increase the opportunity for public involvement, and for the disclosure of otherwise hidden interests and values. They also reinforce legal norms and a legal discourse, often at the expense of a scientific one.

Aside from the "trade-offs" in standard setting, there are some fundamental contradictions in the process. Standard setting is an instrument of industry and national competition, but it demands a high degree of their co-operation. Standard setting involves the public and private sectors in a symbiotic relationship, but the debates about standards focus on the alternatives of regulation versus deregulation. Standards restrain trade and encourage the use of predatory trade practices, but they are also both necessary and desirable in facilitating trade and in supporting free market economies. Standard setting involves a high degree of co-ordination among organizations and countries, but it occurs in an anarchistic mix of government and non-governmental groups. Standards are intrinsically related to the protection of public health and safety, but decisions about accepting them are taken for many other reasons. Standard setting involves the resolution of many mixed disputes, but the procedures for the adequate resolution of truth-seeking and justice-seeking disputes are very different. These contradictions are endemic to standard setting, and they will not be resolved by a modest program of reform. They comprise the infrastructure of standard setting and, we suspect, mandated science as well. It is time return to the more general discussion of mandated science and identify the contribution of our research to understanding it.

Chapter Eight

# Mandated Science

In chapter one, we defined mandated science as that which is used for the purposes of making policy. In this definition, we included original research, both that commissioned for the express purpose of regulatory decision making and the academic studies relied upon by policy makers. We included the evaluations of expert committees and inquiries, and the formal adjudication of science-related issues by regulatory agencies and tribunals. Finally, we included studies commissioned by expert committees, government departments or regulatory agencies to further their aims. The defining characteristic of mandated science was that it was either produced and/or interpreted for the purposes of public policy. Our contention was that the mandate to produce scientific conclusions to support policy decisions had a significant impact upon the conduct of scientific activity, and the interpretation of scientific information.

In arriving at a definition of mandated science, we avoided making a comparison between mandated and conventional science. In fact, we argued that mandated science could include academic research, if it was interpreted and used for the making of public policy. In this case, scientific work was mandated because its audience was policy makers, not scientists, and they would be likely to draw implications from the research that were beyond the original intentions of the researcher. Other forms of mandated science differ from the conventional practice of academic science to the extent that the goals of policy making influenced their design and scope, or that the primary activity of scientific work was the evaluation of research for its policy-related conclusions. We cannot ascertain from our research how many of the characteristics of mandated science also pertain to more conventional work. What is different about mandated science is that the relationships between science, values, public policy and economics are acknowledged and explicit within it. We think this is significant in its own right.

Standard setting was chosen as the example of mandated science because the specific character of standards highlights some of the relationships of interest to us. As we have argued in the last chapter, the relationships among science and economics, law and values are relatively easy to identify in the case of standards.

Moreover, standards are seen to be the science of the regulatory process, and the terminology used in standard setting illustrates some of the more prominant features of how language is used in mandated science. In this chapter, we want to generalize from the study of standard setting to mandated science. We realize that some caution is necessary; standard setting is only one example of mandated science, and the controversies we have chosen represent a small proportion of those in which science and politics play a role. Nonetheless, our own study has been broad enough to give us a number of different examples of mandated science and some points of comparison between them. As well, there is now a growing literature on science and decision making. We have not cited this literature, believing that the empiricial material in our study of standard setting is sufficiently demanding of the reader, but it has informed our understanding of mandated science.

## The Character of Mandated Science

If standard setting is typical, there are a number of characteristics of mandated science. First, science and policy were integrated into a single process in mandated science. With the exception of the small proportion of academic research used by policy makers, questions about public policy were openly recognized as influential in the conduct of research and its interpretation in mandated science. The scientists in our study had little control over the agenda and timing of their scientific assessments, relying instead on industry and government to determine their priorities for research. They had difficulties gaining access to the information they needed, and their activities were subject to scrutiny from those outside the scientific community. They were often forced to adapt to political and legal norms to defend their scientific conclusions. They wrote primarily for a non-scientific audience, yet their reports were supposed to be scientific ones.

Second, mandated science was usually evaluative in orientation. Evaluation of the work of others is always part of the scientific enterprise, but in mandated science, the proportion of evaluation to original research was reversed. Mandated science consisted mainly of the evaluation of the research of others, and only a small proportion of mandated science occurred within a laboratory setting. As well, evaluation of the scientific literature was the end product of mandated science, while in more conventional academic settings, a literature review is usually only the first step in a program of research.

Third, we found mandated science to be neither innovative nor dedicated to open-ended investigation. One scientist suggested to us that mandated science was not science, but "screening". By this he meant that mandated science – his reference was to toxicology, but he also included the work of the expert committees in standard setting – had very limited objectives, few of which involved furthering the scientific enterprise. We, too, found that open-ended investigation of natural phenomena was rare in studies commissioned expressly for the purpose of regulation or public policy, and more particularly for standard setting. For example, every innovative approach taken by either the university scientists in the Toronto lead controversy and the Swedish scientists in the pentachlorophenol case resulted in the rejection of their conclusions by regulators or expert committees. We think that the requirement that conclusions be justified to many different audiences, or in a courtroom, was what discouraged innovation and unorthodox approaches. Lawyers and policy makers sought the assurance of well-established approaches in using scientific work for making public policy. In turn, the lack of possibilities for investigative and innovative work discouraged participation from academic scientists in mandated science, and the experience of participation was frustrating for them.

Fourth, we observed that mandated scientists took an unusually critical perspective on the studies they were reviewing. Indeed, negative evaluations of specific studies, including studies published in recognized academic journals, were very common. We believe that this was a product of the burden placed upon the scientists who conducted the assessments. Since their evaluations were being used to support legal actions and to evaluate regulations, these scientists were likely to exercise great caution in their interpretation of the material at hand. In the face of inconclusive data or conflicting interpretations of it, scientists often chose to identify problems in the conduct of the research or to refer to scientific uncertainty. We also found that advocate groups portrayed science and scientists as conservative, and as unsupportive of stringent regulations. If their assessments were correct, it was perhaps because statements about scientific uncertainty – which were neither conservative nor liberal in a scientific context – could easily be used in mandated science as a reason for government inaction on a particular chemical.

Fifth, we have observed that there were no established and articulated procedures for carrying out mandated science, although formal methods for risk assessment have been adopted in some countries. The procedures varied from committee to committee and from country to country. Sometimes studies or expert evaluations were published, and sometimes not. Sometimes expert committees were used, but these expert committees occasionally conducted their work as if they were regulatory tribunals and heard submissions from political, advocate and interest groups. Sometimes expert committees relied upon conventional peer review; other times studies were simply reviewed by regulators, government officials or in public hearings. Indeed, the variation in the conduct of mandated science was striking, especially given its reliance upon a standardized scientific methodology to determine the merit of the studies being evaluated.

Sixth, we have noted that much of the data being reviewed in mandated science were proprietary, the private property of the companies which sought registration of a particular product. We believe that the conception of science as an intrinsically public enterprise is altered when access to its information is restricted. This is particularly true if the procedures, such as peer review and publication, that are used in conventional science to secure its public status are not applied in mandated science. Our view runs contrary to an opinion we heard expressed often: mandated science was the most public of all scientific activity, because some proportion of the material was reviewed through public hearings with due process and wide scale public participation. We believe that the claim that mandated science is public has merit only to the extent that the proprietary nature of much of the data does not interfere with its accessibility, and only in those relatively infrequent instances when full-scale public hearings are used.

Finally, the scientists who participated in mandated science faced many constraints that would probably have been unacceptable in a more conventional scientific context. For example, the information used in mandated science took many forms and the procedures for its submission varied even from time to time before the same expert committee. The scientists integrated different types of studies, on a variety of very different topics, into a single scientific report. They transposed scientific conclusions from one context to another without appearing to take account of how these conclusions would be understood. They conducted activities that combined scientific and regulatory judgements, but presented their work as if it were simply scientific. They were chosen because their scientific expertise was supposed to protect them from interest group considerations, but functioned in an environment where interests were being negotiated. They had to deal with the fact that mandated science was directed towards closure, towards the production of conclusions that would support decisions taken in another sphere of activity, government or economic relations.

We conclude that the result of these constraints was often a cursory review of the academic literature by policy makers and regulators. On one hand, such regulators had access to studies designed for the purposes of regulation, submitted for their review, and amenable to the type of conclusions they required. They did not need to search the literature widely in order to do their job. On the other hand, resources were not made available for a serious examination of a wide-ranging academic literature, and little scope existed for discussing the implications of studies that did not easily lend themselves to conclusions suitable for time-pressed regulators. For example, on several occasions we were shown a computer print-out of an academic literature search as the only evidence that the literature had been reviewed. No resources had been allocated to its analysis. Indeed, because of the time pressures and the lack of resources for assessment, an informal "old boys" network seemed to be responsible for much of the dissemination of scientific information not directly connected to the regulatory issue at hand. It seemed that at least some regulatory scientists went to conferences with other regulators to find out what – if anything – was worthy of attention in the academic literature.

In summary, if one were to apply the norms of conventional science to the activities of mandated science, then it might be dismissed as unscientific, and as interest-laden. If, on the other hand, one applied the norms of a courtroom to the testimony of scientists, then the particular nature of the scientific enterprise would be lost. We would argue that only a scientist experienced in mandated science could have withstood the scrutiny applied in an expert hearing or a court. Other scientists – whose work was judged excellent by the standards of academic science – usually failed to defend their work adequately when questioned by a skilled lawyer. For example, in the Toronto lead case, the debate in the environmental assessment board was not a scientific debate. Yet scientists – highly reputable scientists – were involved, and their studies were conducted in a credible scientific manner. In the pentachlorophenol case, the Swedish scientists were highly reputable in their field, but they too were treated as lacking credibility. The companies' lawyer in the Toronto lead controversy concluded that scientists are individuals who fail to admit to the interests and political convictions that biased their work. His word for scientists was "scoundrels". Our conclusion is different. It is that mandated science requires its procedures and resources for scientific assessment.

## Questions Arising from the Study of Mandated Science

The study of mandated science gives rise to some fundamental questions about the place of values in a scientific assessment. First, is it possible in mandated science for scientists not to be advocates of one position or another in an interest group conflict? Does scientific information necessarily "line up" on one side or another of a conflict? Obviously the same questions might be asked of conventional science. In mandated science, however, these questions are highlighted, because there is little evidence of the successful participation in mandated science of scientists who are unaffiliated with interest or advocate groups.

Second, if conventional science and scientists fare poorly in mandated science, does this mean that conventional science has no place in mandated science, or that the scientific assessments conducted for mandated science are fundamentally flawed by the standards of conventional science? Is a relatively dispassionate scientific inquiry possible in mandated science? Obviously different criteria will be used in mandated science to judge the merit and applicability of a research study. Do these criteria so alter the practice of science so as to render it unfamiliar to those in the scientific community? In other words, can mandated science include a relatively dispassionate inquiry?

Third, mandated science relies heavily upon an idealistic picture of the scientific enterprise. To what extent is it necessary to defend scientific work in idealistic terms in order to protect the credibility of scientific assessments? Is the ideal science a product of situations in which political chemicals have been created? Does the picture of the scientific enterprise in ideal terms lead to overly rational procedures for mandated science? Is the interest in institutional design, particularly in conjunction with risk assessment, an attempt to take the uncertainty and political conflicts out of a process that uses uncertain information regularly and that is politically controversial by its very nature?

Fourth, mention has been made about the uses of the phrase "scientific uncertainty" in mandated science. To what extent is the science in mandated science actually uncertain, or are other types of questions about its scientific status more relevant? Can different types of scientific uncertainty – each with different implications for mandated science – be distinguished? What are the implications of different types of uncertainty for mandated science, and how might one restrict the use of the phrase "scientific uncertainty" for purely strategic purposes?

Finally, mandated science often involves public hearings, with the resulting introduction of legal norms for the conduct of scientific and regulatory assessments. The norms and values of science are different from those of the legal process. These two norms cannot simply be combined, for they impose contradictory obligations upon the participants in mandated science. Is the problem of integrating legal and scientific norms so great that all efforts to resolve it are useless? The study of standard setting raises questions about the efficacy of both courts and expert committees in conducting scientific assessment. Does this mean that both are to be avoided? What alternatives exist or can be designed? We will address each of these questions in turn.

### *Mandated science and advocacy*

As background to examining the relationship between mandated science and advocacy, it is necessary to clarify our approach to the long-standing debate about the relationship between science and values, for advocacy is based on the articulation of values and interests. Our approach has been to rely upon a sociological definition of science, and to consider it as a particular strategy – or group of related strategies – for knowledge-seeking. In choosing this approach, we have put aside the philosophical question of whether the strategy is ever, or can ever be completely successful in its own terms.

In the last chapter, we added an analytical distinction between truth-seeking and justice-seeking disputes. We said that there was an important difference in orientation in the two kinds of disputes, and that the procedures for their resolution were fundamentally at odds. We suggested that mandated science was best characterized by mixed disputes. Relying on Thibault and Walker's formulation, then, our question is not whether science can be combined with advocacy. It is about the role of advocacy in truth-seeking aspects of mixed disputes. Specifically, what happened to the academically-oriented scientists in our case studies, and did the laypublic or advocate groups make a contribution to the truth-seeking aspects of the controversies we studied?

The last question is easier to answer. At the very least, the advocate groups instigated the review of standards in almost every instance. Moreover, it was their particular concerns that set the agenda for the research in both the lead and pentachlorophenol controversies. For example, the research in the Toronto lead controversy was designed to indicate the extent to which the citizen protesters and the people they represented were actually being exposed to and harmed by the lead from the plants in their neighbourhood. It was not a study of the ambient levels of lead in Toronto. In the pentachlorophenol case, one of the major disputes was over the design of the epidemiological studies, for the union was concerned that the problems they identified would not be evident in the study being proposed. Advocacy, then, serves as a catalyst for standards, and particularly for their revision, but it also inserts a range of considerations into a chemical assessment that otherwise might be neglected, and contributes directly to the design of the research.

Do laypeople make a contribution to the truth-seeking activities of mandated science? The answer to this question depends upon what kind of research is being done. In the truth-seeking aspects of toxicological research, there is little scope for the layperson's contribution, although laypeople contribute to the justice-seeking debate about how toxicological findings should be interpreted for the purposes of making regulations or policy. The layperson's contribution to epidemiological research lies in his or her willingness to participate in the studies, and to lay aside privacy or civil rights in the interests of truth-seeking. This is a relatively passive contribution, to be sure, but it should not be ignored.

There is another type of contribution that laypeople might make to truth-seeking. The best illustration is the pentachlorophenol case, where workers sought a site-specific evaluation of risk. Their argument was that the quantitative assessments of risk were misleading unless they took into account the way the chemical was likely to be used. We have noted that this type of risk assessment is rare when risk is being quantified. The actual conditions of exposure, and the contribution of workplace practices, are usually taken into account only in a more informal, and relatively unsystematic way by the government officials who conduct investigations and contribute to standards.

It is now fairly common practice for inquiries in Canada to use the testimony of laypeople as evidence, scientific evidence about the nature and scope of a problem. The difficulty with this type of evidence is that it is often unsystematic, anecdotal and difficult to quantify. Obviously, if such evidence is to be incorporated in a more formal and systematic way into a risk assessment, attention will have to be directed to how to conduct rigorous studies of the relationship between workplace practices and chemical exposure. Were such studies to be conducted, then, the layperson's contribution to truth-seeking would be related to a site-specific evaluation of risk, and it could be combined with the other quantitative estimates of risk.

The final question concerns the activities of the academic scientists in mandated science. We have noted that such scientists tended to be "adopted" by advocate groups and that their credibility suffered as a result. We observed that the university scientists were easily attacked by a skilled lawyer, because they had not hesitated to make recommendations on the basis of the research they had conducted, unlike the expert witnesses hired by the companies. We noted that the perhaps unwitting reference to morality, and the not-very scientific reference to the submissions of Monsanto by the Swedish scientists only served to undermine their claim to being good scientists. Our question is whether the advocacy, or the seeming advocacy of these academic scientists in fact reduced the credibility of their science. Or did their association with advocacy simply provide a skilled lawyer with an argument for dismissing them.

There can be no question that both the university and the Swedish scientists were engaged in truth-seeking, and that within the limits of their disciplines and resources, they conducted studies that were unexceptional from a scientific point of view. At the same time, these scientists were participating in a mixed dispute, where the procedures for truth-seeking and justice-seeking were likely to clash. The problem arising from mixed dispute was only evident, however, in their oral contributions, and when they were speaking to non-scientists about the conduct of their research. It was also most evident when they were forced, by the nature of the problem being investigated, to explore innovative methods of research. What seems to be true is that the problem of combining science and advoacy occurred when scientists and their work were being evaluated outside the scientific community and by non-scientists.

It is this observation that leads us to identify the problem in the relationship between science and advocacy not in the conduct of the research but in its presentation to a non-scientific community. We think that the university scientists' recommendations would not have been made if these scientists had been talking only to their fellow scientists. The Swedish scientists might have defended the release of their data to the press in terms of its impact upon the research, but not on moral grounds, had they been speaking with scientists. The innovations explored by both groups would not have attracted much attention in a scientific debate; either their work would have been reviewed favourably for publication or it would not. We conclude that advocacy and science clashed in the public presentation of science, rather than its conduct. It is ironic that in mandated science, where science and values are combined explicitly, the clash between advocacy and science is most evident.

There is an important qualification to our conclusion. When discussing advoacy, we have referred to the combination of science and values. We believe it is essential to distinguish between science and values, science and interests, and science and bias, even though all these relationships are often described in terms of science and values. In dealing with the question of science and values, we have argued that the particular problems in mandated science of combining science and advocacy arise from the presentation of scientific information to non-scientific audiences. We have put aside the philosophical questions about whether all science is value-imbued, by concentrating on the situation of mandated scientists.

In dealing with the relationship between science and interests, we reiterate that mandated science is characterized by mixed disputes. We have argued that it is very difficult to separate a mixed dispute into its constituent parts, and to conduct truth-seeking and justice-seeking activities separately and sequentially. This is the reason why we have suggested that mandated science cannot rely upon the norms of either science or the legal process, but requires its own procedures for the evaluation of scientific work. In examining the contribution of the scientists in the Toronto lead controversy or the pentachlorophenol case, we have come to believe that there is no necessary contradiction in holding advocate views and practicing credible science. Company scientists did find hazards with particular chemicals, and reported them. In the Toronto lead controversy, the attempt to discredit some scientists on the basis that they expressed an opinion on regulatory options was a legal manoeuvre, not the result of a considered analysis of science. The Swedish scientists were not better or worse scientists because they pointed out the coincidence between the Monsanto submission and the final report of the Australian Commission. Interests came into play most significantly when their scientific material was being presented for debate in a non-scientific context, when funding for specific research was tied to interest group considerations, or when the full disclosure of information was precluded because of the constraints imposed upon expert witnesses by those who hired them.

Some scientific work is biased, in the most explicit sense of the term. Scientists can be "bought"; data can be falsified; lack of disclosure by expert witnesses is sometimes tantamount to deception. It is important to identify when scientists are corrupt, and when they are not. It is important to know how much industry influence is reflected in decisions about standards. It is essential to know when a study might be replicated by a scientist with other views, and when it has been designed intentionally only to illustrate a point of view. Even if one considers all science to be value-imbued, and mandated science to be "interested", it is still necessary to investigate whether scientific work has been corrupted in specific instances.

We have concluded that the statement that all science is value-imbued disguises the differences between particular uses of science and between the activities of individual scientists. It is counter-productive for the study of mandated science. It masks corruption because it fails to distinguish between the normal range of combinations of science and values in judgements about the conduct of research, the role that interests and politics play in mandated science, and the corruption of science and deceptions of some of its practitioners. The statement that science is always deeply infused with interests discredits the scientific enterprise, not because it is false, but because the product of scientific activity need not always reflect the economic or political interests of the researchers or those who fund them. Science is, among other things, a particular strategy for seeking knowledge about some kinds of phenomena. At its best, it is a particularly useful strategy for evaluating chemical hazards, whatever its other limitations.

*Is mandated science dispassionate*

Mandated science raises the question of whether a relatively dispassionate inquiry is possible. On the basis of our research on standard setting, it is possible to answer this question, but the answer is a complex one. In the last chapter, we identified two sciences in standard setting. In fact, we would argue that there are four activities called science in mandated science, and that the potential for relatively dispassionate inquiry is different in each one of them.

One activity called science is the conduct of original research. In mandated science, this includes two types of research: studies from the conventional scientific literature and studies designed and carried out for the sole purpose of regulation. Both types of research are conducted by scientists in scientific settings, and both follow scientifically conventional methodologies. In assessing whether a relatively dispassionate inquiry is possible in mandated science, it is necessary to take account of some subtle differences in these two types of research.

Each type of research has a different audience and purpose. A study designed for the purposes of regulation must be amenable to conclusions that are useful to regulators. The range of questions to be addressed in the research reflects its purpose. This is what was meant when a scientist called mandated science "screening, not science". We have suggested that a study carried out for the purpose of regulation is likely to be conducted in a highly conventional manner, without resort to innovative techniques or untested methodologies. It is unlikely to conclude with a list of questions to be pursued, or by drawing attention to problems in the conduct of the research. A study designed for the purposes of regulation is unlikely to further the cause of science, however much it contributes to a specific scientific assessment.

Mandated scientists contend that the differences between these two types of research have little bearing upon their conclusions. If this is true, a caution is necessary. Many of the questions raised in the conduct of mandated science – questions such as the appropriateness of different animal species for toxicological testing – cannot be addressed adequately by studies designed exclusively for the purposes of regulation. To the extent that resources are lacking for an adequate assessment of more conventional studies, or that mandated scientists fail to examine the conventional literature *in its own terms*, then these questions remain unanswered. These unanswered questions contribute to the creation of political chemicals, which in turn undermines the attempt to generate a relatively dispassionate inquiry about any specific chemical.

The second activity referred to as science is the scientific evaluation of research conducted for the purposes of policy making. Here too, the answer to the question of whether a relatively dispassionate inquiry is possible is a complex one. We have noted that the evaluation normally takes place under constraints that conventional scientists might well find unacceptable. Usually the scientists conducting the evaluation fully control neither their access to data, nor the priorities for their assessment. They have only limited control over when to release their conclusions, since they are pressured to reach decisions as soon as possible. Their resources are very limited, and they are mindful of the economic, trade and social implications of the decisions they make. It is possible to argue that a relatively dispassionate inquiry is possible, even under these circumstances. In evaluating the claim to be dispassionate, however, it is important to distinguish when the practice of this science is so constrained as to make any dispassionate inquiry impossible.

The third meaning of science arises in conjunction with the JMPR or TLV committees. Science in this case is a comparative term; it is a means of indicating the difference between activities which are scientific and those which are not. JMPR is a scientific committee *as compared with* the other Codex organizations. Its recommendations are the science of the Codex process as compared with other types of assessment conducted by Codex. The term science in this third example refers to something that is obviously both scientific and non-scientific in origin, the recommendations of the JMPR. It is openly inclusive of values, and of judgements that are not scientific but regulatory in orientation. Nonetheless, the claim that the JMPR recommendations are scientific ones is also not without merit. In this instance, however, it makes little sense to refer to science as a relatively dispassionate activity.

The fourth use of the term scientific is also illustrated with reference to the JMPR and the TLV committees. In these instances, science is a method or strategy for cordoning off particular activities from the more political and regulatory assessments that follow. Science refers to an arena of activity that is relatively protected from the by-play of interest group negotiations. For example, JMPR members do not acknowledge what they know as regulators in their own country. In this, they are similar to the members of the TLV committee who, despite their apparent connection to industry, conceive of themselves as engaged in scientific activity and as "watchdogs" over industry. Science in this last case refers to the attempt to conduct a relatively dispassionate assessment in the face of conditions that undermine it. Science is used as a means of controlling the inevitable role that interests play in the conduct of regulatory assessments, and as such, few would claim it was dispassionate.

Thus, there is no one answer to the question of whether a relatively dispassionate inquiry is possible in mandated science. The role played by values, interests and national or political interests is different, depending on the way that science is being used. In each instance of mandated science, and with reference to each type of science, the potential for a relatively dispassionate inquiry is different. Nonetheless, there are serious implications to questioning the potential for dispassionate inquiry in any aspect of mandated science. All mandated science takes place in a context where adversarial relations are encouraged. The stakes are high in a regulatory debate; the companies and their lawyers are defending something more than the quality of the research they have commissioned. Little tolerance is exhibited for those who claim to be dispassionate, especially if it can be argued that these scientists have expressed views on topics beyond their scientific expertise or are otherwise associated with advocate groups.

This situation poses a paradox for conventional scientists. As scientists, they are committed to carrying out relatively dispassionate inquiries. As citizens, and as a result of their research, they are often propelled to "take a position" or to make recommendations for regulation. They hear regulatory or government scientists speak extensively about the options for regulation, and offer their opinions freely. They distrust the seemingly neutral expert witnesses used by the companies, who coincidentally reach favourable – but carefully scientific – conclusions from the companies' perspective. At the same time, any association with advocate or interest groups, or attempts to make regulatory recommendations is seen to corrupt their own science and threaten its scientific credibility. It is easiest to resolve the paradox by simply refusing to become involved in any variant of science connected with mandated science.

### *Mandated science as an idealized science*

We have noted that some people believe that mandated science is the most public form of scientific activity. Their argument is based on the contention that the assessments, the procedures by which they are developed and the conclusions from mandated science must all be presented to an audience that is composed of other scientists, industry, regulators and judges, and members of advocate groups. This demands a science that can be easily understood publicly and adequately justified according to several different criteria including public ones. At the same time, we have argued that mandated science cannot be understood as an intrinsically public enterprise, since many assessments take place behind closed doors, since much of the information is considered to be private property and since many of its studies are not peer reviewed.

There is merit to both arguments, and indeed the public aspect of mandated science poses a paradox for its practitioners, for their work is both highly public in the sense of being justified publicly, and less-than-public by the normal canons of science. To resolve this paradox, we suggest that mandated science often relies upon an idealized picture of the scientific enterprise. For example, we found that mandated scientists and regulators often argued that science is value-free, objective, independent of interest group negotiations, and highly rational. They spoke approvingly of the scientific process that invariably produced a solid foundation of information for the use of regulators. They presented this picture of an ideal science, while fully cognizant of the very heavy strictures laid upon scientific inquiry in mandated science. They did not dissemble. They required science to be the kind of enterprise they described, in order to rely upon science in the policy process.

Regulators and others in mandated science used the ideal picture of science to defend the role of science in regulation and proposed regulatory initiatives. In labelling the products of committees such as the JMPR as science, they were able to instill respect for them and to command resources for further scientific assessment. Their experience in regulation suggested that without the designation of this work as science, it was unlikely that such committees would have been established or taken very seriously.

In cordoning off some activities from interest group negotiations and labelling them as scientific, regulators were also able to control some of the overt influence-peddling and interest group negotiations that would have otherwise occurred. For example, because it was a scientific committee, the JMPR could limit its contact with industry representatives when its draft documentation was being prepared. It could close its meetings to observers who had other interests to pursue. It could assert some control over its agenda, which was otherwise determined by non-scientific factors. The ideal picture of science provided a protective shield, and limited the influence of those who sought to dictate to standard setting organizations decisions in their own favour.

There was an obvious distortion in viewing science, especially mandated science, as an ideal, even if it served the purpose of regulators to do so. In mandated science, the ideal picture of the scientific enterprise had another effect. It led to an overly rational picture of the scientific process, and of the relationship between science and policy decisions. Risk assessment was understood, inappropriately, as a step-by-step process that was capable of disentangling scientific and policy questions. The design of government institutions and regulatory processes was seen to resolve the contradictions in mandated science in the relationship between science and public policy.

What happened to this ideal when chemicals became political chemicals? In each of the controversies, it was impossible to distinguish a single process of chemical assessment. In the pentachlorophenol case, even the regulatory chronology was different from the perspective of the different participating departments. In both the case studies, the ideal picture of science also disguised a reality that was only marginally scientific. In the Toronto lead controversy, scientists were engaged in a process of assessment, and the debate purported to be scientific. The scientists dealt with issues that were not amenable to scientific determination – negligence and liability, for example. They explored areas of scientific uncertainty that were indeterminant by the techniques available to them as scientists. They spoke before a judge who had little understanding of scientific research. They were seen to engage in a scientific debate, but in fact they did not.

When political chemicals were being evaluated, the result was often a parody of science and of rational administrative process. To the extent that the debate became a "set piece", it took place on a variety of different "stages", which were chosen by the protagonists to further their aims. The specific attributes of the chemical were ignored. The debate about captan, a political chemical and another case study in our research, provides a good illustration of the point. Although it was reasonable in the assessment of captan to discuss the relevance of using different animal species in toxicological research, the discussion would only have been scientific if information about the effect of captan on particular species was used to resolve the controversy. But in the public assessment of captan, the debate about the choice of species was a general one, seemingly without reference to the types of tumors produced by captan, or the biological effects of exposure to this particular pesticide.

Our conclusion is that an important characteristic of mandated science is that it is capable of being presented to policy makers and the public as a highly rational, objective and relatively value-free enterprise. We contend that these attributes of science are characteristics of an ideal picture of the scientific enterprise. The ideal serves the purpose of regulators, and in some senses, it protects the scientific component of chemical evaluations. It generates a particularly unrealistic picture of scientific activity in light of the real constraints placed upon mandated science, and it makes a parody of administrative process and science when political chemicals are being evaluated.

### *Uncertainty in mandated science*

Uncertainty is a tenet of adequate science, an assumption of its working practitioners. In mandated science, uncertainty caused problems for regulators and those who depended upon science to support their decisions. Uncertainty in mandated science was a problem to be resolved, not a tenet of good research, and it was a means of discrediting a research study. Indeed, in mandated science uncertainty undermined the scientific foundation of public policy decisions and stalled regulatory action, since regulators were aware that commissioning further research seldom resolved scientific uncertainties.

We think that it is important to explore the ramifications of uncertainty in mandated science more fully. It is our contention that several different uses of the phrase "scientific uncertainty" can be identified in mandated science, and that each has different implications for the resolution of the truth-seeking components of the debate. In our study of standard setting, four different meanings of the term scientific uncertainty were identified. One type of uncertainty – characteristic of research on carcinogenesis – was the result of the underdeveloped state of current scientific knowledge. We would suggest that this was authentic scientific uncertainty, because scientists should be able to resolve it, given enough time and resources.

Second, we have referred to the uncertainty faced by the judge in the Toronto lead controversy, and claimed that this was not, strictly speaking, an example of scientific uncertainty. The judge could not determine whether the pollution came from the factories or the highway, because the exposure to lead occurred in a natural, not a laboratory situation. At best, and with many more resources than the judge was ever likely to command, some measure of probability could be developed. We suggests that uncertainty can only be considered to be scientific uncertainty if scientific research is a reasonably accessible method for resolving it. A better term for the judge's dilemma, given the constraints under which he was operating, is practical indeterminism, for his uncertainty had little to do with questions that scientists were likely to be able to answer.

A third kind of uncertainty is common to social science, but it can be found in the case of the Toronto lead controversy as well. This type of uncertainty refers to the problems introduced when particular techniques of science preclude arriving at conclusions that display a high degree of certainty. We call this methodological uncertainty. The epidemiological research debated in the pentachlorophenol case is an example of methodological uncertainty, rather than scientific uncertainty. Most social scientific research is characterized by methodological uncertainty, for it is unlikely that social science techniques could give rise to a level of certainty comparable to toxicological research. Obviously, there is a continuum in research between results that display more or less certainty. In referring to this third type of uncertainty as methodological uncertainty, we are emphasizing that some kinds of research and some research techniques are much less likely than others to produce the levels of certainty often demanded of science or scientists. The uncertainty in these instances is inherent in the type of research or its research methodologies, and it is not amenable to resolution with the simple application of more time or resources.

The fourth type of scientific uncertainty is best described by the use of an image. Scientific knowledge can be compared to a beam of light, which when further observed through refraction of a prism, becomes more complex and variegated. The original observation of the light as white light is neither scientifically inadequate, nor misleading. Neither is the observation of the more complicated refracted light. The capacity of scientific work to result in ever more complex, and ambiguous conclusions can be described as a form of scientific uncertainty. If science is described in this manner, we should understand that the uncertainty resides – as it did in the case of our image of the light – in deciding where the observations should be made. It does not reside in the science.

The existence of four types of uncertainty has important ramifications for the study of mandated science. In mandated science, scientists are often asked by regulators to provide conclusions they are not ready to provide. This is what we have called genuine scientific uncertainty, and it is readily understandable to policy makers and scientists. Some scientists will be more comfortable than others in dealing with this type of scientific uncertainty, and these scientists are more likely to contribute to mandated science, for time contraints often require conclusions under conditions of genuine scientific uncertainty.

The second type of uncertainty – practical indeterminism – results in inappropriate expectations about science and scientists. If within any realistic projection, neither the time nor the resources are likely to be made available to resolve scientific uncertainty, scientists will inevitably fail policy makers, unless they are prepared to compromise their own standards in presenting their science for review. For example, the scientists in the Toronto lead controversy could have claimed more certainty than their methods, time or resources permitted. They did not, believing that only an admission of uncertainty was compatible with the practice of an adequate science. When the government officials' and court's expectations for scientific certainty were not fulfilled, the university scientists were blamed, and their credibility was questioned. We think that the expectations of science and of scientists in mandated science are often unrealistic.

The third type of uncertainty – the inherent limitations of some observational and measurement techniques to display high levels of certainty – is intrinsic to particular kinds of research. Not all natural or social phenomena are conducive to observation by scientific techniques that display relatively high levels of certainty, even with the most advanced analytical techniques. We believe that much is lost when research characterized by methodological uncertainty does not withstand scrutiny in a regulatory setting, for social scientists and others who use relatively uncertain methodologies, have much to contribute to public policy.

The fourth type of uncertainty is the most difficult one to deal with in the context of mandated science. If scientific research is designed to be an open-ended exploration of the characteristics of natural phenomena, the result is an ever more complex and indeterminant picture of that reality. This is not the kind of science required by mandated science, but it contributes to the stock of knowledge that is necessary if chemical hazards are to be identified adequately. The scientist who designed his or her research to produce relatively unambiguous answers to a narrow range of questions can legitimately speak of the relative conclusiveness of the resulting research. The scientist committed to open-ended investigation has a more difficult time in mandated science, especially if he or she is subject to cross-examination. Labelling this problem as scientific uncertainty obscures the differences between mandated and conventional science, and the limitations of mandated science in terms of the type of scientific conclusions it can produce. It also discourages innovation and more comprehensive scientific work in the conduct of mandated science.

## The Debates in Mandated Science

In dealing with both science and uncertainty, we have placed emphasis on the role of language in mandated science. For example, our assertion was that a number of quite different activities are called science in mandated science. We conclude that the potential for dispassionate inquiry is quite different in each case and that, in at least one instance, science was used as a strategic label to protect the activities of such groups as the JMPR from overt corruption by interest groups. We also have described four different meanings of the term scientific uncertainty, arguing that each type of uncertainty (labelled as scientific uncertainty) had different implications for the conduct of mandated science and for expectations of science and scientists in it. Again, we placed stress upon the strategic uses of language and, in particular, the phrase scientific uncertainty in mandated science.

We believe that mandated science is often characterized by the strategic use of particular terms. The strategic use of language is both a product and a characteristic of mandated science. On one hand, it reflects the influence of interest groups in mandated science, and decisions by lawyers and others about how to present information in a manner most condusive to their particular interests. For example, the argument that science is value-imbued would have quite a different significance were it to come from a philosopher or a lawyer representing his or her clients' interests. For the philosopher, the argument reflects some assertions about the nature of science; for the lawyer, it is intended to discredit the activities of particular scientists or the conclusions of particular studies. The strategic use of language is a product of the mixed nature of the disputes in mandated science, and of the role that interests play in them. It is one of the reasons why mixed disputes are not easily separated into their constituent truth-seeking and justice-seeking components.

On the other hand, the requirement to produce conclusions for public policy generates the strategic use of science itself. The best way to illustrate the point is to provide a hypothetical debate between an epidemiologist and toxicologist discussing the certainty of their respective research conclusions. In this debate, scientists from each discipline are engaged in truth-seeking, and neither is particularly influenced by interests other than scientific ones. The toxicologist, in the hypothetical debate, argues that his science is more certain than epidemiology. With rigorous attention to experimental conditions and the use of control groups, he can control the exact conditions of exposure and can measure the harm with some precision. His statements are certain, he suggests, because the methodology of his discipline facilitates a high degree of experimental control. What he knows, he knows with certainty about the link between exposure and harm.

The epidemiologist replies. "I know," she says, "that the techniques employed by my discipline preclude the kind of careful control you have over experimental conditions in identifying the relationship between exposure and harm. In most instances, I measure harm some time after the exposure has taken place, and often I am dependent upon the recall of my subjects, a notoriously inadequate source of information. I also lack much information that I need in order to estimate exposure accurately, for while I can know whether a chemical was used in an industrial process, I cannot measure, in retrospect, the exact amount of the chemical to which each worker was exposed. To deal with these problems, my discipline has introduced some techniques of its own to foster certainty, and we rely heavily upon probability statistics for our estimates of exposure and harm".

"That said," she continues, "I am not convinced that your science is any more certain than my own. Once your experimental studies are completed, you must engage in a series of extrapolations to arrive at any conclusions about the toxicology of particular chemicals to humans. For some of these extrapolations, you have developed fairly well-established conventions about how they should be done. These are just conventions, of course. The fact that they are widely-accepted does not mean that they are scientifically certain in the sense of being valid. Indeed, you also depend upon probability statistics to arrive at your quantitative estimates of risk. Without information on how a chemical is metabolized within the human body, you lack the conditions for scientific certainty in the application of your conclusions to human populations. In our discipline, we cannot be as confident as you are about the certainty of our experimental results, but we can be at least as confident – probably more – about the application of our findings to human populations".

Both of the scientists in our hypothesized debate would claim some degree of certainty for their conclusions, and both would rely upon the phrase "scientific uncertainty" to discuss the problems with the disciplinary approach taken by the other. In this instance, the meaning of scientific uncertainty lies, in part, in how it is interpreted by those who draw implications from it. As well meaning as our hypothetical scientists are, under the pressures of mandated science, they, too, will engage in a strategic use of language to protect the credibility of their own work, and possibly even to undermine the credibility of others. They, too, have interests which are manifest in a strategic use of language, but in this instance their interests are related to their expertise and disciplines.

Two other aspects of language are important in mandated science. The first is related to the use of technical terms in the debate on chemical hazards. We noted that terms such as "ADI" and "good agricultural practice" had a different meaning in Codex than they did in everyday speech. Confusion occurred for those not aware of the Codex system of standard setting, for if anyone believed that acceptable daily intake related to the actual pesticide residues that people ate, then the ADI seemed to provide a guarantee of public safety. If delegates to CCPR believed that "good agricultural practice" reflected the farming practices in their own countries, they mistakenly applied this standard to situations for which it was inappropriate. The knowledgeable members of the CCPR were not misled by these terms, nor did they use them intentionally in a manner designed to mislead others. The terms themselves were deceptive, because they had very different meanings in regulation than they did in public discussions of safety.

This point is especially important if we assume – as we do – that decisions about standards should also be public discussions of safety. To the extent that the terminology used in standard setting is subject to two different interpretations – an insider's and a public interpretation – then the debate is skewed from both perspectives. For the insiders, a number of irrelevent considerations are brought to bear as a result of the public's misunderstanding of what is intended to be a very limited assessment of toxicological data. For example, inexperienced participants have often concluded that the lack of data about human exposure meant that a chemical was only proven dangerous to animals. Newspaper columnists have mocked studies that relied upon massive doses of a chemical in an animal study, not understanding why these large doses were administered or what was appropriately concluded from them.

For the public, the double meaning of critical terminology has meant that it easily missed the point in the scientific debate. As a result, risks were overestimated and underestimated, and the credibility of public interest groups suffered. Even when members of the public were not misled by the terminology, the use of technical terms with ordinary meanings in the regulatory debate was likely to be intimidating. The belief was common that participation in the public debate about chemical hazards should be left to the expert groups who were most comfortable with the technical language.

The second aspect of communication in mandated science refers to the method of presenting information and argument. In the discipline of communication, this is called discourse. Information is presented in different ways in each type of discourse, and arguments that are effective in one discourse are often ineffectual in another. Participants in mandated science attested to the fact that there were three discourses at play in mandated science, a scientific, a legal and a regulatory discourse. The frustration expressed by the Swedish scientists at the hands of regulators was partly a result of their inexperience in the conventions of regulatory discourse. By contrast, the expert witnesses used by the companies in the Toronto lead controversy were particularly well attuned to the conventions of a legal discourse, and their testimony was more credible than that of the university scientists as a result. Just as we found it useful to create a hypothetical debate in discussing the strategic use of language by scientists, so in this instance, three hypothetical statements made by scientists will best illustrate the point. Our point is that these statements have different significance in regulatory and scientific discourses.

Each of the following statements is unexceptional from a scientific point of view. "We cannot know with certainty that exposure to lead causes these specific injuries". "We have no evidence that the pesticide captan is dangerous to human populations". "We cannot pinpoint the source of the contaminant precisely". In a scientific discourse, the first statement means only that research is still to be conducted or that the results are incomplete or inconclusive. In regulatory discourse, this same statement will often be used to suggest that regulatory action is premature or unnecessary, because a problem has not yet been identified clearly.

The second statement, "We have no evidence that the pesticide captan is dangerous to human populations" is not hypothetical, but is taken from the discussions about the final report of the expert committee on captan. In the scientific discourse about captan, this statement is not very interesting and in the context of the expert committee report, this statement was accurate. Since only toxicological data existed about captan at the time that this statement was made, and since toxicological experiments use animal populations, there could have been no *scientific evidence* that captan was actually dangerous to humans. All that was available to the expert committee were extrapolations from animal studies by scientists who differed in their interpretation of the data. When this second statement was included in the final report of a consultative committee on captan, and when the report was read by people with limited knowledge about the limitations of toxicological research, this statement implied a degree of knowledge about captan that was not available. In a regulatory discourse, it implied that captan was not very dangerous to human populations.

In a scientific discourse, the third statement that, "We cannot pinpoint the source of the contamination precisely" reflects the methodological constraints of conducting research in a natural, rather than a laboratory setting. It means that measurements of probability should be used to obtain an estimate of the relative contributions of various possible sources of contamination. In a regulatory discourse, it might mean that more resources should be made available for research, but in the absence of such resources, that one should be careful about allocating responsibility for the high levels of contamination. In a legal discourse, however, the third statement was sufficient to relieve individual companies from the legal responsibility for any harm resulting from lead emissions from their factories.

In other words, what seems unexceptional from a scientific perspective in each of these three statements is no longer unexceptional when the same statements are made in regulatory or legal discourses. The statements are similar, but the implications drawn from them are not. The point is that statements taking meaning from one discourse have different implications in the others. In mandated science, this applies most particularly to scientific statements that are made, somewhat blindly, as if they had no other possible interpretation. It is important to reiterate that the primary audience for mandated science is not scientists, not even in the case of expert committees. The scientific reports produced by expert committees are designed to be read by people who are not scientists. On the basis of our experience in expert committees, and interviews with scientists, we have concluded that experts regularly take the views of policy makers and regulators into account in writing their reports and arriving at its recommendations. If such scientists are "blind" with respect to the conflicting interpretations of their statements in scientific and regulatory communities, it is either an ill-informed or a willful blindness.

It may be that mandated science has its own discourse, and that the use of specific terms and the interpretation of particular statements are easily understood within the discourse of mandated science by its more experienced participants. We have seen evidence that an unwritten collection of rules govern how information should be presented, and how it will be evaluated in mandated science. The more experienced participants do not seem to fall victim to the same attacks as the university scientists in the Toronto lead controversy or the Swedish scientists in the pentachlorophenol case. Somehow – probably with reference to an unarticulated code of conduct – they manage to maintain some level of credibility as scientists and as participants in the regulatory process.

If we are correct, the point is an important one. On one hand, it is a strong argument for paying attention to the particular norms of mandated science, and for attempting to articulate them for a wider audience. Those who are not experienced in mandated science need to know the "rules" of its discourse if they are not to be excluded from participation in it. On the other hand, it is an argument why even research from academic journals should be considered to be altered when it is used in mandated science. In effect, if particular norms exist in mandated science, then studies from academic journals will be interpreted according to these norms, and differently from their interpretation in a scientific discourse. If we are right, different criteria for their assessment, and different methods of drawing out their implications will be used when a published study is transposed into the world of mandated science.

## Conclusions from the Study of Mandated Science

The study of mandated science explores the role of science and scientists in the making of public policy. We have concluded that scientific and policy issues are so closely entwined that it is impossible to separate them. Procedures, such as those adopted for risk assessment in several countries, have been designed to segment the scientific and regulatory assessment of chemical hazards. These procedures assist the regulators, but they do not resolve the difficulties in the relationship between science and public policy. Too often, the procedures represent an overly rational view of the policy process, and they are the result of viewing science in idealistic terms. The reality of mandated science contradicts the ideal of an objective, value-neutral science. The relationships between science and values – or even between science and interests – in mandated science are much more complex than even we originally envisioned.

The example of standard setting was particularly useful in highlighting several dimensions of mandated science. In that example, the terms "values" and "interests" did not encompass the complexity of the trade and economic relations sustained by standards. These relationships included issues connected with national security, as well as with import and export trade. The relationships among countries, and more particularly between the developed and third world were influenced by standards. Scientific assessment was often used for purposes that have little connection with health, safety or scientific knowledge in standard setting. Our conclusion is that it is necessary to deal with the political and economic implications of mandated science, and not simply to view these implications in terms of the costs imposed by the adoption of stringent standards.

Recently, in dealing with chemical hazards, mandated science has become associated with risk assessment. Risk assessment is based primarily on toxicological data. For intuitive reasons, and because of the particular type of information it provides about the human consequences of chemical exposure, we believe that epidemiological research should be given higher priority in the evaluation of risk. This would mean designing risk assessment procedures that are attuned to the methodological constraints of epidemiological research.

As well, two new types of research should be commissioned to complement current methods of risk assessment. One deals with the way chemicals are used and it involves an assessment of work practices and an examination of the actual conditions in which chemical exposure occurs. How often do farmworkers enter fields that have just been sprayed with pesticides, and why do they do so? Is fruit being washed? What is actually done about spray tanks that leak industrial chemicals into the workplace? In the case of pentachlorophenol, an attempt was made to use codes of good practice as a substitute for a site-specific risk assessment. Regulators argued that it was more useful to change the attitudes of those using pentachlorophenol than to incorporate changing data about the actual conditions of exposure to pentachlorophenol in their assessments. We believe that codes of good practice are voluntary and superficial, and they do not really address the problem of how chemicals are used.

As well, the mathematical estimations of risk used today depend mainly on toxicological data, and as such, they represent a partial picture of chemical hazards. We think that it is necessary to complement these particular mathematical estimations of risk with a different type of risk assessment. We were impressed by the observation in standard setting of how few times anyone conducted a systematic investigation of the effects of altering a standard. Rarely were studies conducted and used in standard setting to see which estimates and assumptions were borne out. We believe that this type of research was conspicuously absent from mandated science, in part because decisions about standards are neither frequently nor systematically revised. Our conclusion is that resources should be allocated for the purposes of evaluating the estimates and assumptions used in standard setting, especially since decisions are often made on the basis of incomplete or inconclusive information.

There are also good reasons to seek an assessment that combines the best attributes of a scientific discussion with the legal norms of fairness, due process and natural justice. We have argued that the combination is not, and cannot be an easy one. In our research, we found that neither the importation of legal discourse into a scientific assessment, nor the transposition of legal norms into mandated science resolved the problems of mixed disputes. The legal norms undermined the scientific debate, but their use in mandated science did not result in any apparent gain in fairness. We believe that the search for fairness in mandated science will demand considerably more innovation than has been demonstrated by experiments with combining truth-seeking and justice-seeking procedures, such as in the science court or mediation techniques. Other methods are required, and their development requires attention to problems other than administrative procedure and institutional design. In the past, the study of mandated science has tended to focus on scientific controversies, partly in order to resolve problems of legal or institutional design that gave rise to controversy. Perhaps it is time to focus on some less controversial decisions, to determine whether their participants – members of the public, scientists, regulators, industry – are satisfied with the result. The most successful experiences in mandated science should be sought out to determine how mixed disputes have been resolved, and how both fairness and adequate science have been achieved.

Is there another way to approach mandated science in general and standard setting in particular? The more radical options have not been evaluated properly, so we cannot know. It is time to put these options on the table, to study and debate them thoroughly. What would it mean to set up government laboratories and require that all studies of chemical hazards be conducted by them? How would the companies' rights with respect to new product development be protected in such a situation? Is there any way to debate the merit of studies used for the purposes of regulation in the open scientific literature without violating the rights of the industries that develop the products? How would money to support such research be collected? From industry which now pays the cost of testing or from the general public? Is there any way to increase the influence of such groups as IARC and IPCS, the international organizations that take a broader perspective than Codex? We do not know the answers to these questions, and we cannot know them until we explore the implications of each option.

We have found that mandated science is not well understood, and that it is characterized by an ideal picture of the scientific enterprise, by reliance upon overly rational procedures of risk assessment, by an insufficient understanding of the difficulties of resolving mixed disputes and by an almost naive perception that scientific statements have the same meaning, regardless of who interprets them. We have found the discussion of the relationship between science and values to be too limited, and the reliance upon institutional procedures to handle the economic, trade and interest group issues in mandated science to be highly problematic. We have argued that mandated science should be studied in its own right, and that it already has some norms that should be extended and articulated to deal with the constraints imposed by a policy mandate upon scientists and scientific work.

The creation of standards is a difficult and complex task. The practice of mandated science is equally so. Very much that is already in place will remain, whatever decisions are now made. Institutions change slowly, and in response to pressures that have little relevance to some of the issues that we have discussed – namely, the adequacy of scientific assessments and their fairness in both procedural and substantive terms. What we will not concede is that we should have confidence in mandated science as it now stands, even if many of the decisions emerging from it are themselves sound. Nor will we support a quest for new procedures for mandated science, for standard setting, or for chemical assessement, if such a quest is based on an inadequate and ideal picture of either mandated science or science.

# Notes

## Chapter One

[1] See, Liora Salter, "Fairness in the Canadian Inquiry Process", *Fairness in Environment Assessment*, (Edmonton: Canadian Institute for Resource Law, 1983).
[2] For a discussion of the difficulties encountered by scientists who must participate, at the same time, in both a scientific and a regulatory discourse, see, Liora Salter, "Science and Peer Review: The Canadian Standard-Setting Experience", *Science, Technology & Human Values*, Vol. 10, Issue 4, (Fall 1985), pp. 37–46.
[3] W. W. Lowrance, *Of Acceptable Risk: Science and the Determination of Safety*, (Los Altos, California: William Kaufmann, Inc., 1976).
[4] See, for example, the proposals for a Science Court outlined in the report of the Task Force of the Presidential Advisory Group on Anticipated Advances in Science and Technology, "The Science Court Experiment: An Interim Report", *Science*, (August 1976), pp. 653–656; W. D. Ruckelhaus, "Risk in a Free Society", *Risk Analysis*, Vol. 4, (1984), pp. 157–162; and, W. W. Lowrance, op. cit..
[5] See, Sheila Jasanoff, *Risk Management and Political Culture: A Comparative Study of Science in the Policy Context*, (New York: Russell Sage Foundation, 1986).
[6] A. Kantrowitz, "The Science Court Experiment: Criticisms and Responses", *Bulletin of the Atomic Scientists*, (April 1977).
[7] B. Latour and S. Woolgar, *Laboratory Life: The Social Construction of Scientific Facts*, (London and Beverly Hills: Sage, 1979); also, Bruno Latour, "Give Me a Laboratory and I Will Raise the World", in Karin Knorr-Cetina and Michael Mulkay, (eds.), *Science Observed*, (London and Beverly Hills: Sage, 1983), pp. 141–170; and, Steve Woolgar, (ed.), "Laboratory Studies", *Social Studies of Science*, Vol. 12, No. 4, (1982), pp. 481–558.
[8] Karin Knorr-Cetina, *The Manufacture of Knowledge: An Essay on the Constructivist and Contextual Nature of Science*, (Oxford: Pergamon Press, 1981).
[9] Among them, Michael Lynch, *Art and Artifacts in Laboratory Science*, (London: Routledge and Kegan Paul, 1982); Doug McKegney, *Decisions, Consequences and Public Explanations: The Relationship Between Research Activity and Formal Knowledge in an Ecological Laboratory*, Ph. D. Dissertation, Department of Communication, Simon Fraser University, Burnaby, B.C., 1982; and, Trevor Pinch, "The Sun-Set: The Presentation of Certainty in Scientific Life", *Social Studies of Science* , Vol. 11, No. 1, pp. 63–93.
[10] The literature is extensive, comprehending most of the contemporary writing in the sociology of science. Representative works, in addition to those previously cited, include: Barry Barnes, *Scientific Knowledge and Sociological Theory*, (London: Routledge and Kegan Paul, 1974); Barry Barnes, *Interests and the Growth of Knowledge*, (London: Routledge and Kegan Paul,

1977); Barry Barnes, *T.S. Kuhn and Social Science*, (London: Macmillan, 1982); David Bloor, *Knowledge and Social Imagery*, (London: Routledge and Kegan Paul, 1976); Augustine Brannigan, *The Social Basis of Scientific Discoveries*, (Cambridge and New York: Cambridge University Press, 1981); H. M. Collins and Trevor Pinch, *Frames of Meaning: The Social Construction of Extraordinary Science*, (London: Routledge and Kegan Paul, 1982); Michael Mulkay, *Science and the Sociology of Knowledge*, (London: George Allen and Unwin, 1979); and Steven Yearley, *Science and Sociological Practice*, (London: Open University Press, 1984).

11 Of the recent work in this area, Michael Mulkay is most explicit. See, G. Nigel Gilbert and Michael Mulkay, *Opening Pandora's Box: A Sociological Analysis of Scientists' Discourse*, (Cambridge and New York: Cambridge University Press, 1984); Michael Mulkay, *The Word and the World: Explorations in the Form of Sociological Analysis*, (London: George Allen and Unwin, 1985). See also, Michael Mulkay, Jonathan Potter and Steven Yearley, "Why an Analysis of Scientific Discourse is Needed", *Science Observed*, Karin Knorr-Cetina and Michael Mulkay, (eds.), (London and Beverly Hills: Sage, 1983), pp. 171–203, as well as other studies in this collection. Although it seems redundant to cite, T. S. Kuhn, *The Structure of Scientific Revolutions* (second enlarged edition), (Chicago: University of Chicago Press, 1962), has attained something like the status of Genesis in its account of the creation and legitimation of scientific knowledge.

12 For a discussion of this process, see Arie Rip, "Legitimations of Science in a Changing World", *Wissenschaftssprache und Gesellschaft*, ed. Theo Bungarten (Hamburg: Edition Akademion, 1986), pp. 133–48.

13 See, for example, the study by the U.S. National Academy of Sciences entitled, *Risk Assessment in the Federal Government*, (Washington, D.C.: National Academy Press, 1983), which goes through an excellent list of points in risk assessments at which value laden assumptions must be made. An earlier book, *Boundaries of Analysis* by Harold Feiveson, Frank W. Sinden and Robert H. Socolow, eds., (Cambridge, Mass.: Ballinger Pub. Co., 1976), provides a useful critique of the uses of science to justify water resource projects. A number of books provide useful case study glimpses of the uses of science in environmental controversies: for three examples, note Robert Crandall and Lester Lave's book *The Scientific Basis of Health and Safety Regulation*, (Washington, D.C.: The Brooking Institution, 1981) which addresses several regulatory cases from the perspective of scientists, economists, and administrators; Arthur Vander's *Nutrition, Stress and Toxic Chemicals*, (Ann Arbor: University of Michigan Press, 1981) which provides a valuable scientist's perspective on the use of scientific information in several controversial health regulation issues; and Robert Bartlett's *The Reserve Mining Controversy*, (Bloomington: Indiana University Press, 1980) which focusses on the treatment of science in the debate over disposal of mining wastes into Lake Superior. Finally, Ted Greenwood's *Knowledge and Discretion in Government Regulation*, (New York: Praeger, 1984), provides a good analysis of the use of science in the regulatory process itself.

14 See, for example, A. M. Weinberg, "Science and Trans-Science", *Minerva*, Vol. 10, (1982), pp. 209–222; Science Council of Canada, "Review and Recommendations: Science and the Legal Process", in *Regulating the Regulators*,(Ottawa: Science Council of Canada, 1983); and, M. R. Wessel, *Science and Conscience*, (New York: Columbia University Press, 1980).

15 Liora Salter and William Leiss, *Consultation in the Assessment and Registration of Pesticides,* Canada, Department of Agriculture, 1984.

16 See William Leiss, *The Risk Management Process,* Canada, Department of Agriculture, October 1985.

17 See, Liora Salter, "Observations on the Politics of Risk Assessment: The Captan Case", *Canadian Public Policy*, Vol. XI, No. 1, (March 1985), pp. 64–76.

## Chapter Two

[1] Government of Canada, Department of Agriculture, Alachlor Review Board (1986–87), proceedings in progress.
[2] For a comprehensive overview of developments related to the asbestos standard, see, Paul Brodeur, "The Annals of Law (Asbestos – Parts I to IV), *The New Yorker*, 10 June 1985, pp. 49–101; 17 June 1985, pp. 45–111; 24 June 1985, pp. 37–77; and 1 July 1985, pp. 36–80.
[3] See, for example, William J. Baumol and W. E. Oates, "The Use of Standards and Pricing for the Protection of the Environment", *Swedish Journal of Economics*, **73**, 1981, pp. 42–54; Paul L. Joskow and Roger G. Noll, "Regulation in Theory and Practice: An Overview", *Studies in Public Regulation*, Gary Fromm (ed.), (Cambridge, Mass.: MIT Press, 1981), pp. 1–65; and A.V. Kneese and C.L. Schultze, *Pollution, Prices and Public Policy* , (Washington: Brookings Institution, 1975).
[4] See, for example, Roger G. Noll and Bruce M. Owen (eds.), *The Political Economy of Deregulation* , (Washington: American Enterprise Institute for Public Policy Research, 1983); also, Sam Peltzman, "Current Developments in the Economics of Regulation", *Studies in Public Regulation*, Gary Fromm (ed.), (Cambridge, Mass.: MIT Press, 1981), pp. 371–384.

## Chapter Three

[1] W. G. Frederick, "The Birth of the ACGIH Threshold Limit Values Committee and Its Influence on the Development of Industrial Hygiene," *1968 ACGIH Transactions*, p. 40. (The full title of the *Transactions* is *Transactions of the Thirtieth Annual Meeting of the American Conference of Governmental Industrial Hygienists*. Hereinafter, I shall use *19-- ACGIH Transactions*.
[2] Ibid.
[3] M. Bowditch, "In Setting Threshold Limits," *1944 ACGIH Transactions*, p. 29.
[4] The identities of a number of individuals whose comments appear in this work have not been provided in order to ensure their anonymity. Discussions were conducted in a full and detailed manner. As a consequence of the not for attribution basis, in each case when interview material is used, information is provided in the text to identify the position of the person involved and, where necessary, the factual basis for the assertion.
[5] "Report of the TLV Committee," *1964 ACGIH Transactions*, p. 111.
[6] *1965 ACGIH Transactions*, p. 125.
[7] ACGIH Membership brochure; no date.
[8] "Proposed Budget: July 1984 to June 1985" (mimeo).
[9] "1984 TLV Committee Report"; mimeo circulated at Annual Meeting.
[10] Based on information provided in the ACGIH – TLV Booklets.
[11] Observation made at the ACGIH Annual Meeting, 1984–85.
[12] *1983-84 TLV Booklet*, p. 3.
[13] Ibid., pp. 3–4.
[14] Ibid., pp. 5–6.
[15] Ibid., pp. 8–9.
[16] Ibid., Appendix A, pp. 41–47.
[17] *1983-84 TLV Booklet*, p. 3.
[18] Ibid., p. 2.
[19] Observation made at the AIHA Educational Seminar 3, 1984–85.
[20] Ibid.
[21] P. Caplan as reported in *1977 ACGIH Transactions* p. 44.

[22] The correspondence of a current and long standing member of the ACGIH - TLV Committee was provided to us.
[23] Observation made at the TLV Committee Meeting, 1984.

## Chapter Four

[1] Secretariat of the Joint FAO/WHO Food Standards Programme, Codex Alimentarius Commission, *Codex Alimentarius Commission: Procedural Manual*, 5th ed. (Rome: FAO/WHO, 1981), p. 3.
[2] D. G. Chapman, *The Codex Alimentarius Commission: General Overview and Historical Perspective*, (unpublished document, not dated), p. 4.
[3] Ibid.
[4] Secretariat of the Joint FAO/WHO Food Standards Programme, Codex Alimentarius Commission, op. cit., p. 21.
[5] Ibid., p. 44.
[6] Secretariat of the Joint FAO/WHO Food Standards Programme, Codex Alimentarius Commission, *Report of the Fifteenth Session of the Joint FAO/WHO Codex Alimentarius Commission*, (Rome: FAO/WHO, ALINORM 83/43, September 1983).
[7] Ibid.
[8] Kenneth C. Walker, "International Aspects of Pesticides," *Industrial Production and Formulation of Pesticides in Developing Countries, Vol 1: General Principles and Formulation of Pesticides*, United Nations Industrial Development Organization, (New York: United Nations, 1972), p. 20.
[9] Ibid., pp. 26–27.
[10] Working paper of the Codex Committee on Pesticide Residues, ALINORM 79/24 CX/PR 78/5, March 1978.
[11] Joint FAO/WHO Food Standards Programme, Codex Alimentarius Commission, *Report of the Seventeenth Session of the Codex Committee on Pesticide Residues*, (Rome: FAO/WHO, 1986).
[12] Memorandum to: Codex Contact Points – Participants in the 13th Session of CCPR, from: J. M. Stalker, Associate Director, Technical Services Section, Pesticides Division.
[13] Working paper of the Codex Committee on Pesticide Residues, op. cit..
[14] Working paper of the Joint Management Committee on Pesticide Residues (JMPR), 1978.
[15] *Pesticide Residues in Food: Report of the 1983 Joint FAO/WHO Meeting of Experts*, FAO Plant Production and Protection Paper No. 56, (Rome: FAO, 1984), pp. 65–67.
[16] Joint FAO/WHO Food Standards Programme, Codex Alimentarius Commission, *Report of the Tenth Session of the Codex Committee on Pesticide Residues*, (Rome: FAO/WHO), 1986.
[17] Working paper of the Joint Management Committee on Pesticide Residues (JMPR), 1978.
[18] "*Maximum Residue Limit*" (MRL) is the maximum concentration for a pesticide residue resulting from the use of a pesticide according to good agricultural practice that is recommended by the Codex Alimentarius Commission to be legally permitted or recognized as acceptable in or on a food, agricultural commodity or animal feed. The concentration is expressed in milligrams of pesticide residue per kilogram of the commodity." As defined in: *Recommended National Regulatory Practices to Facilitate Acceptance and Use of Codex Limits for Pesticide Residues in Food*, Working document prepared by the ad hoc Working Group on Regulatory Principles, Codex Committee on Pesticide Residues, (Rome: FAO, CX/PR 84/8, April 1984), p. 33.
[19] See, for example: *Reconsideration of Codex Definition of Terms in Light of Definitions Adapted by the Joint FAO/WHO Meeting on Pesticide Residues*, paper prepared by the

Secretariat, (CX/PR 80/21, April 1980). See also: *U.S. Comments to the Ad Hoc Working Group on Regulatory Analysis*, (not dated).

[20] *Pesticide Residues in Food: Report of the 1983 Joint FAO/WHO Meeting of Experts*, op. cit., p. 5.

[21] *Codex Alimentarius, Volume XIII: Codex Maximum Limits for Pesticide Residues*, (Rome: FAO/WHO, CAC/VOL XIII – Ed 1, 1983), p. 7-iv.

[22] Joint FAO/WHO Food Standards Programme, Codex Alimentarius Commission, *Report of the Seventeenth Session of the Codex Committee on Pesticide Residues*, op. cit..

[23] See: Organization for Economic Co-operation and Development, *The OECD and its Chemicals Programme*, August, 1983, (W7299A).

[24] The JMPR is not empowered with the authority to order that specific studies be carried out. While it can request that countries and/or companies provide data on a specific pesticide, compliance with JMPR requests for data is entirely voluntary.

## Chapter Five

[1] Donald A. Chant, Frank A. DeMarco and H. Rocke Robertson, *Report of the Committee to Inquire into and Report upon the Effect on Human Health of Lead from the Environment*, (Toronto: Ministry of Health, October 29, 1974) pp. 42–43.

[2] James F. MacLaren Limited, *National Inventory of Sources and Emissions of Lead (1970)*, (Ottawa: Environment Canada, November 1973), Air Pollution Control Directorate, E.P.S., Internal Report APCD 73-7.

[3] See, Liora Salter, "Science and Peer Review: The Canadian Standard-Setting Experience," *Science Technology & Human Values*, Volume 10, Issue 4, (Fall 1985), pp. 37–46.

[4] Bruce G. Doern, Michael Prince and Garth McNaughton, *Living with Contradictions: Health and Safety Regulation and Implementation in Ontario*, Study No. 5 of the Royal Commission on Matters of Health and Safety Arising from the Use of Asbestos in Ontario, (ISBN: 0-7743-7056-4, February, 1982), p. 2.38.

[5] Ibid., p. 2.41.

[6] G. L. Stopps, *Summary Statement on Air Quality Standards and Criteria*, Paper presented December 19, 1974, to the Environmental Hearing Board, p. 18.

[7] *Review of Air Quality Standards for Lead*, Report submitted with a letter from W. B. Drowley, Executive Director, Air and Land Pollution Control, Ministry of the Environment, to G. W. Moss, Public Health, City Hall, Toronto, dated December 13, 1973.

[8] We have located the regulatory formula by which the standards for dustfall and soil were set, but not the method of conversion from ambient air exposure to blood lead levels.

[9] Other evidence that airborne standards were used as the basis for all other standards is also taken from: M. L. Phillips and H. P. Sanderson, *Report on Air Quality Objectives for Airborne Lead*, (Downsview, Ont.: Atmospheric Chemistry, Criteria and Standards Division, Atmospheric Environment Service, April 9, 1974), Internal Report ARQA-3-74. See Table 10, where assumptions for derivation of blood lead levels are stated.

[10] See, Bruce C. Martin and P. C. Kupa, *The Rationale, Methodology and Administration Used in Ontario to Determine Ambient Air Objectives and Emission Standards*, Paper presented at the 70th Annual Meeting of the Air Pollution Control Association, June 19–24, 1977.

See also: Robert B. Gibson, *Control Orders and Industrial Pollution Abatement in Ontario*, (Toronto: The Canadian Environmental Law Research Foundation, 1983) Note. 37, p. 38 for other references to procedures.

[11] For a description and analysis of this procedure see, Robert C. Gibson, op. cit..

[12] See, Bruce G. Doern, et al, op. cit., p. 2.28. Doern says the asbestos standards were the

ACGIH standards.

[13] During the controversy there were several lawyers involved from the environmental law advocate group. Different lawyers represented different groups at different times. An account of the environmental group's involvement can be found in, *CELA News*, 1973.

[14] Letter dated July 7, 1972, from Ian Roland to M. S. Smith, Director of Legal Services, Ministry of the Environment.

[15] For a discussion of the various control orders issued during and after the controversy, see, Linda McCaffrey, *CELA News*, Vol. 6, No. 1, pp. 20–21.

[16] For a description of these events, see, Wendy C. MacKeigan, *The Case of Lead Pollution in Toronto: Scientific Information and Public Response*, unpublished MA dissertation, University of Toronto, 1975, p. 16.

[17] Refer to Canada Metal Co., et al and MacFarlane (1973), D.L.R. (3d) 161.

[18] Letter dated November 15, 1973.

[19] Press Release dated January 2, 1974.

[20] Ibid..

[21] See, H. R. S. Ryan, "The Trial of Zundel, Freedom of Expression and the Criminal Law," *Criminal Reports* , (Third Series), Vol. 44, Part 4, June 1, 1985, 44 C. R. (3d), pp. 334–351.

[22] In June, 1971, The National Indian Brotherhood, Indian-Eskimo Association, Union of Ontario Indians and Canadian-Indian Centre of Toronto proceeded in the Federal Court of Canada against a decision of the Executive Committee of the Canadian Radio-Television Commission that it was not in the public interest to hold a Hearing into the Indian associations' complaint with respect to the film "The Taming of the Canadian West".

[23] A total of ten studies and reports were produced during and after the controversy by the university scientists, not including their brief to the Environmental Assessment Board and a document entitled, *Lead: The Issues and the Risk – A report for Human Environmental Systems*, (Institute for Environmental Studies, University of Toronto), April 14, 1976, prepared by K. Beatty, A. Howell and S. Staiman.

[24] Letter dated July 10, 1975.

[25] In 1979, the Ministry of the Environment brought charges against Canada Metal Co. Ltd. for violation of EPA 1971, S. O. 1971, c86, Reg. 15, for unlawfully causing or permitting the concentration of a contaminant, namely lead, at a point of impingement exceeding the standard as prescribed in Schedule 1 of the Regulation. The company pleaded guilty on the two charges, but the judge granted a suspended sentence, citing the company as a "good corporate citizen".

[26] See, Liora Salter, "Observations on the Politics of Risk Assessment: The Captan Case," *Canadian Public Policy*, Vol. XI, No. 1, (March 1985), pp. 64–76.

[27] Court proceedings Re: Canada Metal Co. Ltd. and MacFarlane, 1973, (High Court of Justice), 31-1 O. R. (2d), p. 587.

[28] Ibid.

[29] Ibid., pp. 589–90.

[30] Ibid., p. 591.

## Chapter Six

[1] K. R. Rao, (ed.), *Pentachlorophenol: Chemistry, Pharmacology and Environmental Toxicology*, (New York: Plenum Press, 1978), p. 394.

[2] American Conference of Governmental Industrial Hygienists – Committee on Threshold Limit Values, *Documentation of Threshold Limit Values* , (Fourth Edition), (Cincinnati: ACGIH, 1981), p. 323.

[3] "The commercial wood preservative products are often mixtures of tetra and pen-

tachlorophenols with lesser amounts of "other" chlorinated phenols or their salts and also contain toxic byproducts such as chlorinated biphenyl ethers (CBE), dibenzo-p-dioxins (CCD) and dibenzo-p-furans(CDF)." from a paper presented at the Symposium on Cancer in the Workplace, University of British Columbia, Vancouver, May 16-18, 1983 by Dieter Riedel, Ph.D., of Health and Welfare Canada entitled: *Potential Long-Term Health Hazards of Wood Preservatives: Problems and Solution*, p. 8.

[4] The dioxins are the hexa, hepta and octa.

[5] See: Statement of Douglas M. Wilson on "The Deleterious Effects of Chlorophenols to Fish", Regina vs. Cloverdale Paint and Chemicals Ltd., August, 1984.

[6] Daniel P. Cirelli, Project Manager, U.S. Environmental Protection Agency, *Position Document 1*, p. 13.

[7] American Conference of Governmental Industrial Hygienists – Committee on Threshold Limit Values, op. cit..

[8] Theodor D. Sterling, Larry D. Stoffman, David A. Sterling and Gabor Mate, "Health Effects of Chlorophenol Wood Preservatives on Sawmill Workers," *International Journal of Health Services*, Vol. 12, No. 4, (1982), pp. 559–571.

[9] See for example: Donald A. Enarson, Moira Chan-Yeung, Valerie Embree, Robert Wang, Michael Schulzer, *Occupational Exposure to Chlorophenates: Renal, Heptic and other Health Effects*, a report from the Occupational Diseases Research Unit, Respiratory Division, Vancouver General Hospital, Department of Medicine, University of British Columbia, (not dated).

[10] This provincial research council gets less of its money from government, and more from industry, than other provincial research councils.

[11] Union letter dated July 21, 1983, from: J. J. Munro, President, IWA to : K. J. Bennett, Forest Labour Relations Council.

[12] The registration of pentachlorophenol is somewhat complicated. Pentachlorophenol was originally registered as a pesticide by Agriculture Canada only because it was used around farms, and likely to enter the food chain as a result. It was explained to us that, " We can regulate the chemical, but we don't necessarily register it per se. If we wanted to, we could remove all of the wood preservatives from the registration system and put them into a schedule of our act, such that the products could only be used for certain purposes. Uniroyal doesn't have a registration for pentachlorophenol in Canada. They sell to other formulators who have registration numbers. They now have applied for a registration number. If Uniroyal was selling direct to pressure treatment plants, the product would have to be registered."

[13] P. A. Jones, *Chlorophenols and their Impurities in the Canadian Environment*, (EPS 3-EC-81-2), March, 1981.

[14] Ibid. p. xxxiii

[15] See: M. F. Mitchell, H. A. McLeod and J. R. Roberts, *Polychlorinated Dibenzofurans: Criteria for their Effects on Humans and the Environment*, (NRCC No. 22846 of the Environmental Secretariat), 1984, p. 26.

[16] NRCC Associate Committee on Scientific Criteria for Environmental Quality, *Polychlorinated Dibenzo-p-Dioxins: Criteria for their Effects on Man and his Environment*, (NRCC No. 18574 of the Environmental Secretariat), 1981, p. 16.

[17] Health and Welfare Canada, Environment Canada, *Report of the Joint Health and Welfare Canada/Environment Canada Expert Advisory Committee on Dioxins*, November, 1983.

[18] See: Interdepartmental Committee on Toxic Chemicals, Environment Canada, *Dioxins in Canada: The Federal Approach*, December, 1983.

[19] See especially: O. Axelson and L. Sundell, "Herbicide Exposure, Mortality and Tumor Incidence. An Epidemiological Investigation on Swedish Railroad Workers", *Scan. J. Work, Environment, Health*, **11**, (1974), pp. 21–8.

[20] L. Hardell, and A. Sandstrom, "Case Control Study: Soft-tissue Sarcomas and Exposure to Phenoxyacetic Acids or Chlporophenols", *British Journal of Cancer*, **39**, (1979), pp. 711–17; O.

Axelson, L. Sundell, K. Anderson, C. Edling, C. Hogstedt, and H. Kling, "Herbicide Exposure and Tumor Mortality: An Updated Epidemiological Investigationon Swedish Railroad Workers", *Scan. J. of Work, Environment & Health*, **6**, (1980), pp. 73–79; L. Hardell, M. Eriksson, P. Lenner, and E. Lundgren, "Malignant Lymphoma and Exposure to Chemicals Especially organic Solvents, Chlorophenols and Phenoxy Acids: A Case-Control Study", *British Journal of Cancer*, **43**, (1981), pp. 169–76; L. Hardell, "Relation of Soft-tissue Sarcoma, Malignant Lymphoma and Colon Cancer to Phenoxy Acids, Chlorophenols and other Agents", *Scan. J. of Work, Environment & Health*, **7**, (1981), pp. 119–130; M. Eriksson, L. Hardell, N. O. Berg, T. Moller, and O. Axelson, "Soft-tissue Sarcomas and Exposure to Chemical Substances: A Case-Referent Study", *British Journal of Industrial Medicine*, **38**, (1981), pp. 27–33; L. Hardell, and N. O. Bengtsson, "Epidemiological Study of Socio-economic Factors and Clinical Findings in Hodgkin's Disease and Reanalysis of Previous Data Regarding Chemical Exposure", *British Journal of Cancer*, **48**, (1983), pp. 217–225.

[21] D. Coggon, and E. D. Acheson, "Do Phenoxy Herbicides Cause Cancer in Man?", *Lancet*, May 8, 1982, pp. 1057–59; C. M. Bishop and A. H. Jones, "Non-Hodgkin's Lymphoma of the Scalp in Workers Exposed to Dioxins," *Lancet*, August 15, 1981, p. 369; P. A. Honchar and W. E. Halperin, "2,4,5-T, Trichlorophenol and Soft-tissue Sarcoma", *Lancet*, **i**, (1981), pp. 268–269; C. Edling and S. Granstam, "Causes of Death among Lumberjacks – A Pilot Study", *Journal of Occupational Medicine*, **22**, (1980), pp. 403–406; J. .P. Seiler, "The Genetic Toxicology of Phenoxy Acids other than 2,4,5-T", *Mutat Res*, **55**, (1978), pp. 197–226.

[22] USA – EPA hearings on 2,4,5-T and Silvex; 1980. PD 1 published, 43 FR 48443, 10/18/78: PD 2/3 (wood use only) completed and Notice of Determination published in 46 FR 13020, 2/19/81, comment period closed PD 4, 5/13/83.

[23] Correspondence collected by a participant in the debate was made available to us.

[24] Royal Commission on the Use and Effects of Chemical Agents on Australian Personnel in Vietnam, *Final Report*, July, 1985, p. viii–112.

[25] Ibid., p. viii–116.

[26] B. K. Armstrong, "Storm in a Cup of 2,4,5-T", *The Medical Journal of Australia*, Vol. 144, March 17, 1986, pp. 284–85.

[27] See: Health and Welfare Canada, Environment Canada, *Report of the Joint Health and Welfare Canada/Environment Canada Expert Advisory Committee on Dioxins*, November, 1983.

[28] Also see: E. Levy, "The Swedish Studies of Pesticide Exposure and Cancer: A Case Study of Disciplinary and Mandated Science", *Alternatives*, forthcoming.

## Chapter Seven

[1] John Thibault and Laurens Walker, "A Theory of Procedure," *California Law Review*, **66** (May 1978), pp. 541 and 543.

[2] Ibid., p. 557.

[3] Ibid., p. 544–45.

# Index

Acceptable Daily Intake (ADI), 82, 83, 84, 93–96, 163, 177, 203
Advocate and citizen groups, 98, 99, 108, 109, 110, 112, 120, 123, 124, 125, 133, 155, 156, 157, 162, 166, 181, 190, 192
Agent Orange, 18, 122, 132, 150, 151, 152
Alachlor, 29
American Conference of Governmental Industrial Hygienists (ACGIH), 15, 24, 25, 36–66, 97, 98, 103, 125, 134, 137, 161, 162–165, 168, 172, 176, 177, 179
American Industrial Hygiene Association (AIHA), 36, 39, 40, 43, 60, 63, 141
American Standards Association (ASA), 37, 38

Biological and genetic screening, 58, 59, 180, 182
Biological Limit Values, 50, 57, 58, 59, 180
Bureau of Occupational Health and Safety (BOSH), 39, 41

Canadian Standards Association (CSA), 24, 179
Cancer, 12, 18, 30, 52, 84, 134, 143, 144, 158, 199
Cancer policies, 12
CAPTAN, ix, 79, 83, 123, 124, 199, 205
Carcinogens, 11, 50, 84, 153, 177
Ceiling limits, 44, 49, 55, 62, 63
Cigarette smoking, 30, 158
Codes of professional conduct, 37, 182
Codes of ethics, 72
Codes of Good Practice, 29, 32, 57, 73, 88, 138, 145, 147, 207
Codex Alimentarius, 47–73, 80, 170, 176, 177, 196, 203, 208
Codex Committee on Pesticide Residues (CCPR), 15, 24, 25, 67–97, 125, 161, 162–165, 168, 169, 170, 172, 176, 203
Confidentiality, 29
Consensus Standards, 14, 27, 28, 42, 170
Consumer groups and representatives, 25, 69, 70, 75, 137, 162
Courts, role of, 3, 6, 31, 52, 64, 88, 101, 107, 109–113, 115, 116, 119, 125–128, 130, 131, 159, 163, 173
Cross-examination, 120, 129, 130, 167, 172, 173, 175, 201

Data problems, 29, 30, 31
Democratic rights and values, 7, 66, 162, 172, 173
Deregulation, 14, 31, 144, 145, 172, 178–180, 182
Dioxins, 18, 123, 132, 134, 135, 138, 146, 148, 149
Due diligence, 163, 170, 180
Due process, 131, 162, 172, 173, 174, 183, 208
Dumping of pesticides, 92, 170, 171

Economic Interests, 12, 13, 89
Environmental Audits, 102
Epidemiological studies and research, 30, 31, 121, 141, 142, 143, 145, 148, 150, 154, 156, 158, 177, 192, 200, 207
Environmental Protection Agency (EPA) (United States), 16, 25, 65, 103, 118, 136
Epidemiological studies and research, 30, 31, 121, 141, 142, 143, 145, 148, 150, 151, 154, 156, 158, 177, 192, 200, 207
Epidemiologists, 30, 45, 142, 144, 153, 157, 202
Ethics, 43, 58, 78, 180, 183
European Common Market, 16, 53, 67, 86, 87, 88, 89, 151, 157, 161
Excursion limits, 49, 55, 61, 62, 63
Expert Committees, 2, 3, 8, 9, 38, 67, 74, 76,

79, 84, 88, 100, 101, 104, 117, 118, 119, 121, 130, 131, 138, 148, 151, 157, 163, 173, 183, 188, 205
Expert witnesses, 98, 100, 107, 109, 111, 112, 120, 125, 127, 129, 175, 181, 194, 204

Fairness (see also Democratic rights), 7, 23, 177, 183, 208
Food and Agriculture Organization (FAO), 67, 69, 70, 74, 76, 78, 81, 83, 84, 86, 91
Food and Drug Administration (FDA) (United States), 16, 65

GIFAP, 75, 86, 90, 91, 93
Good agricultural practice, 82, 83, 84, 94, 95, 97
Good laboratory practice, 88
Governmental Standards, 27, 28

Harmonization, 28, 69
Hazardous waste, x
Hearings, 16, 98, 120, 130, 131, 190

Industrial Bio-test (IBT) Laboratories, 29
Insurance, 31, 36, 60, 61, 65, 66, 137, 163, 170, 180
Interest groups, 7, 27, 143, 174, 189, 190, 196, 198
International Agency for Research on Cancer (IARC), 16, 68, 88, 89, 208
International Labour Organization (ILO), 54
International Program for Chemical Safety (IPCS), 16, 68, 88, 89, 208
International Standards Organization (ISO), 24, 25, 179

Joint Management Committee on Pesticide Residues (JMPR), 67–97, 157, 176, 177, 196, 198

Knorr-Cetina, Karin, 11

Labour unions, 46, 47, 59, 60, 75, 99, 132, 139–145, 154, 156, 177
Language, 7, 8, 10, 13, 93–97, 165, 187, 202, 203, 204
Latour, Bruno, 11
Lawyer(s) as participants, 18, 99, 105, 106, 107, 108, 109, 111, 120, 124, 129, 130, 167, 173, 188, 190, 193, 202
Lead, 17, 18, 19, 98–131, 165–167, 173, 181, 190, 194, 200, 204
Liability, 6, 60, 108, 130, 165, 173, 198
Litigation, 64
Lowrance, W.W., 10

Media, 110, 115, 116, 123, 152, 157
Mixed disputes, 175, 176, 178, 184, 191, 193, 202, 208

National Cancer Institute (NCI) (United States), 45
National Institute for Occupational Safety and Health (NIOSH), 40, 41, 45, 63, 103, 139
National Reseach Council (NRC), 101, 148, 154
Negligence, 6, 108, 130, 173, 198
Non-tariff barriers to trade, 90, 92, 170

Ontario Environmental Assessment Board, 98, 119–122, 167, 173, 190
Occupational Health and Safety Administration (OSHA) (United States), 25, 36, 40, 41, 42, 45, 48, 53, 63, 65, 103, 161
Organization for Economic Cooperation and Development (OECD), 16, 68, 88

Peer review, xii, 2, 3, 88, 118, 129, 143, 174, 188, 189
Pentachlorophenol, 18, 19, 122, 132–159, 165–167, 177, 190, 192, 194, 198, 200, 207
Performance Standards, 14, 32, 33, 104, 127, 180–182
Political chemical, 99, 113, 122–125, 132, 133, 153, 167, 177, 198, 199
Pollution, 21, 22, 33, 34, 37, 89, 98, 99, 101, 103, 105, 108, 112, 114, 116, 118, 122, 169
Prescriptive Standards, 14, 32, 33, 103, 104, 128, 180–182
Private sector, 19, 179, 180
Product Specific Standards, 26
Proprietary data, 6, 29, 31, 88, 179, 188, 189, 197
Public debate, 34, 96, 124, 162, 167, 204
Public hearings, 16, 29, 101, 113, 115, 125, 131, 138, 154, 155, 161, 162, 173, 188, 189

Public Inquiries, xi
Public involvement, 97, 100, 101, 162, 165, 189
Public perceptions of acceptable risk, 161
Public sector, 19, 179, 180

Registration Standards, 26, 27
Regulatory Standards, 12, 14, 26, 27, 28, 31, 32, 36, 41, 42, 55, 56, 60, 64, 104, 163, 173, 178, 179
Risk(s), 21, 34, 55, 57–59, 168, 175, 176, 177, 181, 182, 192, 204
Risk Assessment, 10, 12, 16, 21, 156, 168, 175, 176–178, 188, 192, 198, 206, 207
Risk Evaluation, 12, 155
Risk Management, 10, 12, 172

Safety factor, 82, 83, 95
Science advisory councils, 101, 115, 117
Science court, 10, 178, 208
Science and values, 5, 9, 11, 74, 176, 178, 191, 193, 194, 206, 208
Science Council of Canada, 101
Science, values, and public policy, 10, 11, 17, 34, 156, 183, 186
Scientific methodologies, 2, 5
Scientific norms, 4, 7, 9, 11
Scientific uncertainty, 7, 8, 117, 166, 188, 191, 198, 199–201, 203
Self regulation, 27
Short-term exposure limit (STEL), 43, 44, 49, 55, 61, 62, 63, 64
Standardization, 13, 28, 62, 69, 170

Thibault, John, 174, 175, 178, 191
Third World, 54, 68, 70, 72, 91, 207
Thompson, Andrew, xi
Toxicological assessments, 55, 56
Toxicological data, 49, 50, 63, 64, 82, 83, 94, 169, 177, 204, 205, 207
Toxicological testing, 28, 55, 56, 64, 82, 84, 94, 95, 157, 192, 200
Toxicology, 78, 108, 150, 157, 202
Trade and commerce, 13, 24, 68, 69, 70, 72, 74, 76, 80, 89–93, 162, 168, 170, 171, 184, 207
Trade associations, 70, 71, 75, 89, 120, 166
2, 3, 7, 8 – TCDD, 134, 146
2, 4 – D, 149, 153
2, 4, 5 – T, 132, 138, 150, 151, 153, 157

Underdeveloped countries, 70, 72, 77, 84, 86, 90, 91
Unions (see also Labour unions), 46, 60, 139–145, 154, 156

Value debate, 11, 183
Value-free (science as value-free), 5, 9, 198, 199
Voluntary Standards, 12, 14, 27, 28, 32, 41, 68, 163

Walker, Laurens, 174, 175, 178, 191
Walsh-Healy Act, 41, 42
Woolgar, Steve, 11
Workers' compensation, 53, 54, 65, 66, 132, 137, 138, 140, 166
Workers' Compensation Board, 17, 54
World Health Organization (WHO), 67, 68, 70, 74, 76, 77, 78, 80, 84, 86, 89